Zoran Majkic

Differential Equations for Elementary Particles: Beyond Duality

Zoran Majkic

Differential Equations for Elementary Particles: Beyond Duality

LAP LAMBERT Academic Publishing

Impressum / Imprint

Bibliografische Information der Deutschen Nationalbibliothek: Die Deutsche Nationalbibliothek verzeichnet diese Publikation in der Deutschen Nationalbibliografie; detaillierte bibliografische Daten sind im Internet über http://dnb.d-nb.de abrufbar. Alle in diesem Buch genannten Marken und Produktnamen unterliegen warenzeichen-, marken- oder patentrechtlichem Schutz bzw. sind Warenzeichen oder eingetragene Warenzeichen der jeweiligen Inhaber. Die Wiedergabe von Marken, Produktnamen, Gebrauchsnamen, Handelsnamen, Warenbezeichnungen u.s.w. in diesem Werk berechtigt auch ohne besondere Kennzeichnung nicht zu der Annahme, dass solche Namen im Sinne der Warenzeichen- und Markenschutzgesetzgebung als frei zu betrachten wären und daher von jedermann benutzt werden dürften.

Bibliographic information published by the Deutsche Nationalbibliothek: The Deutsche Nationalbibliothek lists this publication in the Deutsche Nationalbibliografie; detailed bibliographic data are available in the Internet at http://dnb.d-nb.de. Any brand names and product names mentioned in this book are subject to trademark, brand or patent protection and are trademarks or registered trademarks of their respective holders. The use of brand names, product names, common names, trade names, product descriptions etc. even without a particular marking in this works is in no way to be construed to mean that such names may be regarded as unrestricted in respect of trademark and brand protection legislation and could thus be used by anyone.

Coverbild / Cover image: www.ingimage.com

Verlag / Publisher:
LAP LAMBERT Academic Publishing
ist ein Imprint der / is a trademark of
AV Akademikerverlag GmbH & Co. KG
Heinrich-Böcking-Str. 6-8, 66121 Saarbrücken, Deutschland / Germany
Email: info@lap-publishing.com

Herstellung: siehe letzte Seite /
Printed at: see last page
ISBN: 978-3-659-40142-8

Differential Equations for Elementary Particles: Beyond Duality

Zoran Majkić

majk.1234@yahoo.com,
http://zoranmajkic.webs.com/

0.1 Preface

This theory of particles is a new constructive approach, alternative to the string theory, where the particles are not points but 3-dimensional conservative distributions of matter with a finite volumes in each fixed time instance during their propagation. Thus, it avoids the infinitary problems of the inverse square low for gravitational and electric forces, and can be used as a basis for the Einstein's unification theory. Their description by differential equations is done in a particular Minkowski's 4-dimensional time-space. The basic particles described in this framework have common basic properties for all kinds of elementary particles, and additional properties that distinguish them (as electric charge, spin, etc..) need additional space-time structures, that can be done, for example, by introducing new compacted space dimensions of very small size wrapped up in one way or another.

In contrast to conventional quantum field theory, which makes gravity impossible, this theory of particles requires gravity. Thus, it is fully integrated with the theory of relativity, and makes complete the quantum theory in sense of Einstein, by overcoming the Copenhagen interpretation. The partial differential equations describe the relativistic phenomena of shrinking of lengths, and, generally, the dependence of particle's geometrical form (its internal matter-distribution) on its velocity and acceleration. A number of well-known physical principles are obtained as derived results of this theory, and are consolidated by 10 detailed examples.

The plan of this work is the following:

The Chapter 2 defines the principal theoretic framework: In Section 2.1 are presented the partial differential equations [20] for the wave-packets in the Minkowski 4-dimensional spaces. In particular we considered the stationary cases for the matter wave-packets, by considering that in all cases, the geometric form (the matter distribution Φ) of a particle in a given time-instance t depends on the particular boundary conditions for these differential equations as well.

In Section 2.2 is presented the energy and special relativity effects for the wave-packets in the Minkowski 4-dimensional spaces [18, 21]. In Example 2 is presented the stationary case for a propagation of a particle, while the unstationary cases are considered in the Examples 3 and 4.

In Section 2.3 the external "field's" forces that interfere with the particles are based on the collisions between them and a particular kind of "field's" particles with rest mass equal to zero (bosons): it is possible because the mass-particles (or excited massless particles) have a wave-packet's distribution Φ with a finite space volume, and, consequently, they may mutually collide with another particles (as bosons of this "field"). In this Section are presented two examples as well: One-dimensional particle in a box problem, and 3rd Bohr postulate for the electrons, both derived from differential equations for the elementary particles.

The Chapter 3 is dedicated to other important results of this new theory. In Section 3.1 we introduced the Lagrangian for the complex wave-packets in the Minkowski 4-dimensional spaces [19] of the massless elementary particles. Differently from common wave-packets in the Euclidian 3-dimensional space, the mathematical expressions for these matter's wave-packet are complex functions, that represent two complemen-

2

tary information about elementary particles: their distribution of a matter and their phase. Then we show how the Lagrangian is determined by the time changing of the phase on the particle's trajectory, and we derive the Fermat's principle. Finally, these results are applied in order to derive the Snell's low for the refraction of light.

In Section 3.2 we consider the relationship of our differential equations with the Klein-Gordon and Schrödinger equations for the elementary particles[21]. In this paper we show that the Schrödinger equation does not describe the propagation of a single wave-packet of an elementary particle but to the coherent stream of particles.

Because of that it has only a statistical meaning that can be applied to the coherent stream of particles in order to compute the quantistic energy levels, for example in "particle in box" problem, for which is demonstrated that the statistical solution obtained by Schrödinger equation is equal to the solution obtained directly from the new differential equations of propagation of a single particle.

In Section 3.3 are used the new differential equations for the elementary particles in order to derive the Young's low and explain the results of Double-slit experiments.

Differently from common wave-packets in the Euclidian 3-dimensional space, the mathematical expressions for the matter's wave-packet are complex functions, and in Section 3.4 we explain what is the reason for that: the fact that with these mathematical expressions we represent two complementary information about elementary particles, their distribution of a matter and their trajectory.

The last two Sections in Chapter 4 are dedicated to the considerations about ERP paradox and Einstein open question about the incompleteness of the quantum mechanics theory and Copenhagen interpretation.

3

Notation convention

Minkowski space [11], where time is imaginary and Euclidean three dimensions real, here we define the basic time-space four mutually orthogonal vectors $e_j, 0 \leq j \leq 3$, by the following matrix:

$$\begin{pmatrix} e_0 \\ e_1 \\ e_2 \\ e_3 \end{pmatrix} = \begin{pmatrix} 1 & 0 & 0 & 0 \\ 0 & i & 0 & 0 \\ 0 & 0 & i & 0 \\ 0 & 0 & 0 & i \end{pmatrix}, \text{ with imaginary number } i = \sqrt{-1}.$$

Here is some other notations:

- a vector of position in this space-time 4-dimensional system, w.r.t. a given referential coordinate system, is given by
  $\vec{r_4} = cte_0 + xe_1 + ye_2 + ze_3 = cte_0 + \vec{r}$;

- $\vec{r}_T(t) = x(t)e_1 + y(t)e_2 + z(t)e_3$ is the vector that lies on the Euclidean 3-dimensional particle's trajectory, dependent on time t;

- The 4-dimensional velocity of a given material point (tangent on its trajectory) in this Minkowski space is then defined by
  $\vec{v_4} = \frac{d\vec{r}}{dt} = ce_0 + v_x e_1 + v_y e_2 + v_z e_3 = ce_0 + \vec{v}$,
  where $\vec{v} = \frac{\partial}{\partial t}\vec{r}_T(t) = v_x e_1 + v_y e_2 + v_z e_3$ is the standard definition of the velocity in the 3-dimensional Euclidean space;

- The trajectory of a material point that moves in the Minkowski 4-dimensional space by a velocity $\vec{v_4}$ is defined by the *unitary tangent* vector of this trajectory
  $\vec{\tau} = \frac{\vec{v_4}}{|\vec{v_4}|} = \frac{\vec{v_4}}{c\sqrt{1-\beta^2}} = \frac{1}{\sqrt{1-\beta^2}}e_0 + \frac{v_x}{c\sqrt{1-\beta^2}}e_1 + \frac{v_y}{c\sqrt{1-\beta^2}}e_2 + \frac{v_z}{c\sqrt{1-\beta^2}}e_3$;

- $E_0\vec{\tau} = mc^2 e_0 + cmv_x e_1 + cmv_y e_2 + cmv_z e_3 = Ee_0 + c\vec{p}$ is the invariant 4-dimensional energy-momentum vector with $E_0 = m_0 c^2$, where $m = \frac{m_0}{\sqrt{1-\beta^2}}$
  is the relativistic mass for a given velocity v of this particle, $E = mc^2$ is its total relativistic energy, and $\vec{p} = m\vec{v}$ is its 3-dimensional momentum;

- 4-dimensional angular wavenumber and momentum vector $\vec{p_4}$, in this four-dimensional time-space are given by
  $\vec{k_4} = k_t e_0 + k_x e_1 + k_y e_2 + k_z e_3 = \frac{\omega}{c}e_0 + \vec{k}$, and $\vec{p_4} = \hbar\vec{k_4}$.
  ω is the angular frequency and $\vec{k} = k_x e_1 + k_y e_2 + k_z e_3$, with $k^2 = |\vec{k}|^2 = k_x^2 + k_y^2 + k_z^2$, is the spatial component of the angular wavenumber vector: $k_x = \frac{2\pi}{\lambda_x}$, $k_y = \frac{2\pi}{\lambda_y}$, $k_z = \frac{2\pi}{\lambda_z}$, where $\lambda_x, \lambda_y, \lambda_z$ are spatial wavelengths w.r.t the axes x, y and z respectively;

- $\nabla = e_1 \frac{\partial}{\partial x} + e_2 \frac{\partial}{\partial y} + e_3 \frac{\partial}{\partial z}$ is the gradient, so that the Laplacian is defined by
  $\triangle = -\nabla^2 = \frac{\partial^2}{\partial x^2} + \frac{\partial^2}{\partial y^2} + \frac{\partial^2}{\partial z^2}$;

- Dirac function:
  $\delta(\vec{r} - \vec{c}t) = \delta(x - c_x t, y - c_y t, z - c_z t) = \delta(x - c_x t)\delta(y - c_y t)\delta(z - c_z t)$.

4

Contents

Chapter 1

Introduction

1.1 "God does not play dice"(A.Einstein)

Physicists usually dichotomize the Theory of Relativity into two parts. The first is the Special Theory of Relativity, which essentially deals with the question of whether rest and motion are relative or absolute, and with the consequences of Einstein's conjecture that they are relative.

The discovery was, although all other velocities are relative, the velocity of light was absolute. The speed of light in a vacuum is the same for all observers, regardless of their relative motion or of the motion of the source of the light. This was learned by a series of Michelson-Morley experiments. The other primary postulate used as a basis for special relativity was that the same laws of physics apply in any and all inertial reference frames. Using these postulates and the Lorenz transform equations to mathematically equate two relative inertial reference frames, Einstein's theory resulted in some very strange and difficult to believe outcomes. Some of these include *shrinking of lengths*, slowing down of time, clocks aging, etc. and warping of the space-time continuum.

The second is the General Theory of Relativity, which primarily applies to particles as they accelerate, particularly due to gravitation, and acts as a radical revision of Newtons theory, predicting important new results for fast-moving and/or very massive bodies. The General Theory of Relativity correctly reproduces all validated predictions of Newtons theory, but expands on our understanding of some of the key principles. Newtonian physics had previously hypothesis that gravity operated through empty space, but the theory lacked explanatory power as far as how the distance and mass of a given object could be transmitted through space. General relativity irons out this paradox, for it shows that objects continue to move in a straight line in time-space, but we observe the motion as acceleration because of the curved nature of time-space.

The General Theory of Relativity demonstrates that time is linked, or related, to matter and space, and thus the dimensions of time, space, and matter constitute what we would call a continuum. They must come into being at precisely the same instant. Time itself cannot exist in the absence of matter and space.

Following Max Planck's quantization of light (the black body radiation), Albert Einstein interpreted Planck's quantum to be photons, particles of light, and proposed that the energy E of a photon is proportional to its frequency ν, $E = h\nu$ where h is Planck's constant, one of the first signs of wave-particle duality. Since energy and momentum are related in the same way as frequency and wavenumber in special relativity, it followed that the momentum p of a photon is proportional to its wavenumber $k_0 = \frac{2\pi}{\lambda}$, i.e., $p = \hbar k_0$, where $\hbar = \frac{h}{2\pi}$ and λ is the wave-length.

Louis de Broglie [4] hypothesized that this is true for all particles, even particles such as electrons. Assuming that the waves travel roughly along classical paths, he showed that they form standing waves for certain discrete frequencies. These correspond to discrete energy levels, which reproduced the old quantum condition.

Following up on these ideas, Schrödinger decided to find a proper wave equation for the electron. He was guided by William R. Hamilton's analogy between mechanics and optics, encoded in the observation that the zero-wavelength limit of optics resembles a mechanical system-the trajectories of light rays become sharp tracks which obey Fermat's principle, an analog of the principle of least action. He put together his wave equation and the spectral analysis of hydrogen in a paper in 1926. The paper was enthusiastically endorsed by Einstein, who saw the matter-waves as an intuitive depiction of nature, as opposed to Heisenberg's matrix mechanics, which he considered overly formal. The Schrödinger equation details the behavior of Ψ but says nothing of its nature. Schrödinger tried to interpret it as a charge density in his fourth paper, but he was unsuccessfully. In 1926, Just a few days after Schrödinger's fourth and final paper was published, Max Born successfully interpreted Ψ as a probability amplitude. Schrödinger, though, always opposed a statistical or probabilistic approach, with its associated discontinuities (much like Einstein, who believed that quantum mechanics was a *statistical approximation* to an underlying deterministic theory) and never reconciled with the Copenhagen interpretation.

The Copenhagen interpretation is an currently dominant interpretation of quantum mechanics. A key feature of quantum mechanics is that the state of every particle is described by a wavefunction Ψ, which is a mathematical representation used to calculate the probability for it to be found in a location or a state of motion. According to this interpretation, the act of measurement causes the calculated set of probabilities to "collapse" to the value defined by the measurement. This feature of the mathematical representations is known as wavefunction collapse. There is no definitive statement of the Copenhagen Interpretation since it consists of the views developed by a number of scientists and philosophers at the turn of the 20th century. Thus, there are a number of ideas that have been associated with the Copenhagen interpretation:

1. A system is completely described by a wavefunction Ψ, which represents an observer's knowledge of the system. (Heisenberg)

2. The description of nature is essentially probabilistic. The probability of an event is related to the square of the amplitude of the wavefunction related to it. (Born rule, due to Max Born)

3. Heisenberg's uncertainty principle states the fact that it is not possible to know the values of all of the properties of the system at the same time; those properties that are not known with precision must be described by probabilities.

4. Complementarity principle: matter exhibits a wave-particle duality. An experiment can show the particle-like properties of matter, or wave-like properties, but not both at the same time.(Niels Bohr)

5. The correspondence principle of Bohr and Heisenberg: the quantum mechanical description of large systems should closely approximate the classical description.

Einstein never accepted such a quantum mechanics as a "real" and complete theory, struggling to the end of his life for an interpretation that could comply with relativity without complying with the Heisenberg Uncertainty Principle. As he once said: "God does not play dice", skeptically referring to the Copenhagen Interpretation of quantum mechanics which says there exists no objective physical reality other than that which is revealed through measurement and observation. Einstein believed that quantum mechanics was a statistical approximation to an underlying deterministic theory and never reconciled with the Copenhagen interpretation.

In this manuscript we will show that Einstein was right, and we will develop this underlying deterministic theory, that is, a real deterministic theory for elementary particles in the microcosms, as complement to his General relativity theory in macrocosms.

The current theories did not explain in which way an elementary particle, that propagates with a given velocity with respect to the referential system of an observer, is shortened (for this observer) in the direction of propagation, and in which way the matter of elementary particles appears as a property of time-space.

Differently from the superstring theory, where elementary particles were treated as vibrating wires (called strings or cords) instead of pointlike entities without internal structure as suggested by the so-called Standard Model, here we consider the massive particles (fermions) as three-dimensional distribution of matter that propagates in time, such that at each instance of time they occupy a finite volume different from zero. Only massless particles (bosons) can be approximated as pointlike entities with Dirac distribution function, in particular stationary situations.

Then we will show that Schrödinger equations for particles are only *statistical* consequences obtained for the *stream of particles* of the same type (stream of photons, for example). As a consequence, Max Born probabilistic interpretation for solutions of Schrödinger equations is correct, and, for any given single particle, it describes the probability of its position from a coherent statistical point of view. Thus, there is no any *collapse* for any given particle as presumed by the Copenhagen interpretation.

Here we obtain the differential equations for wave-packets of any elementary particle with all values that are deterministic (not probabilistic), that particle has in each time instance: position, velocity, momentum, masses and total energy. Moreover, we explain in which way, the masses, momentum and energy are derivable from the four-dimensional time-space matter distribution (geometry) of a given particles.

This theory confirms the Einstein thesis that information can never be transmitted faster than the speed of light, and that another interpretations, as for example the possible world interpretation, are without reasonable sense. If we would be able to measure all initial values of a given particle in a given instance of time, we would theoretically be able to determine all its history during an experiment. But it is not possible because of Heisenberg uncertainty principle that is valid for wave-packets, thus the Schrödinger equations will continue to be useful for all experimentations, by adopting its statistical interpretation. What we will avoid is to use them in order to explain the behavior of a single particle in particular situations. The new differential equations for any single elementary particle, presented in this manuscript, can be used more appropriately for such objectives.

Here we will present the fundamental concepts for matter-events in the Minkowski's 4-dimensional time-space, resulting in the stationary wave-packets that propagate in the ordinary 3-dimensional space. Then will be derived the basic differential equations for the matter-distribution during a propagation of the particles. The basic properties as the rest mass, relativistic mass, total energy and momentum will be derived from the corpuscular matter-distribution of stationary particles, and they will explain the effects of the special relativistic theory on the particle's "geometry" as well. Will be introduced the new concepts as "particle's explosion" in particular boundary conditions, and their application to the photon with mass and unstationary virtual particles. After that will be derived the second-order differential equations, where appears the particle's

10

acceleration as well, in order to show how the external forces influence the particle's "geometry" and trajectory. It will be demonstrated that in the stationary cases we obtain a kind of D'Alambert differential equations for the particles. These results are applied to the massive particles: to the one-dimensional particle in a box problem, and to the derivation of the 3rd Bohr postulate for the electrons. Section 5 will be dedicated to the massless particles and to the derivation of the Lagrangian, based directly on the particle's real velocity and not on the phase velocity (used in the current practice). Than it is applied for the derivation of the particle's trajectory, and is derived Snell's low for the refraction of a single photon, based on the original Newton's idea for "corpuscular" light. Particular attention will be dedicated to the demonstration that Klein-Gordon and Schrödinger equations can not define a propagation of a *single* particle, but are statistical consequences of our deterministic quantum theory for the particles, as was an Einstein's opinion. We will demonstrate that a stream of the particles of the same type, energy and momentum, generates a perfect plain wave that explains their statistical wave-effects. Finally, it will be presented the derivation of Young's low from this new quantum theory, in the famous double-slit experiments, and will be shown that Einstein's ideas was correct w.r.t the Copenhagen-interpretation.

Chapter 2

Theory of differential equations

2.1 Matter-events in the four-dimensional time-space

In physics and mathematics, Minkowski space (or Minkowski time-space) is the mathematical setting in which Einstein's theory of special relativity is most conveniently formulated. In this setting the three ordinary dimensions of space are combined with a single dimension of time to form a four-dimensional manifold for representing a time-space.

In theoretical physics, Minkowski space is often contrasted with Euclidean space. While a Euclidean space has only spacelike dimensions, a Minkowski space also has one timelike dimension. We define the basic time-space four mutually orthogonal vectors $e_j, 0 \le j \le 3$, by the following matrix:

$$\begin{pmatrix} e_0 \\ e_1 \\ e_2 \\ e_3 \end{pmatrix} = \begin{pmatrix} 1 & 0 & 0 & 0 \\ 0 & i & 0 & 0 \\ 0 & 0 & i & 0 \\ 0 & 0 & 0 & i \end{pmatrix}, \text{ with imaginary number } i = \sqrt{-1}.$$

Note that the matrix above is a particular case of the Minkowski tensor using a four-dimensional time-space, which combines the real dimension of time with the three imaginary dimensions of space.

Consequently, a vector of position in this space-time 4-dimensional system is given by $\overrightarrow{r_4} = cte_0 + xe_1 + ye_2 + ze_3 = cte_0 + \overrightarrow{r}$, where t is the time (i.e., ct is the timelike component of $\overrightarrow{r_4}$, where c is the velocity of light in the vacuum) and $\overrightarrow{r} = xe_1 + ye_2 + ze_3$ is an ordinary Euclidean vector with x, y, z three spatial coordinates. Its infinitesimal amount is defined by $d\overrightarrow{s} = cdte_0 + dxe_1 + dye_2 + dze_3$, where dt, dx, dy and dz are infinitesimal amounts of time-space dimensions.

Thus, in this 4-dimensional system the time is real while the three orthogonal space coordinates are imaginary. This choice is adopted in order to have that the distance $ds^2 = d\overrightarrow{s} \, d\overrightarrow{s} = (cdt)^2 - dx^2 - dy^2 - dz^2$,
for all local time-space reference systems of observations of quantum events be the positive real value (where space dimensions are limited).

An angular wavenumber vector in this four-dimensional time-space is given by $\overrightarrow{k_4} = k_t e_0 + k_x e_1 + k_y e_2 + k_z e_3$.

In what follows we will denote by $\overrightarrow{k} = k_x e_1 + k_y e_2 + k_z e_3$ the spatial component of the angular wavenumber vector, with $k^2 = |\overrightarrow{k}\,\overrightarrow{k}| = k_x^2 + k_y^2 + k_z^2$, so that $k_4^2 = \overrightarrow{k_4}\overrightarrow{k_4} = k_t^2 - k^2$.

The mutually independent space-components are defined as usual by $k_x = \frac{2\pi}{\lambda_x}$, $k_y = \frac{2\pi}{\lambda_y}$, $k_z = \frac{2\pi}{\lambda_z}$, where $\lambda_x, \lambda_y, \lambda_z$ are spatial wavelengths w.r.t the axes x, y and z respectively, and $\lambda = \frac{2\pi}{k}$ is the (total) spatial wavelength. Let $\omega = 2\pi\nu$ be an angular frequency that depends on the space-components, $\nu = \frac{1}{T}$ with a time period T. Thus, $\lambda_t = cT$ is the time-like wavelength and $k_t = \frac{2\pi}{\lambda_t} = \frac{\omega(k)}{c}$ depends on the space-components in $\overrightarrow{k_4}$, so that it holds that $dk_4 = dk = dk_x dk_y dk_z$, and $-\overrightarrow{k_4}\overrightarrow{r_4} = k_x x + k_y y + k_z z - \omega(k)t$.

Remark: from the relativistic theory we have that for each massive elementary particle (with rest mass m_0 greater than zero) it holds that $\omega(k) = \pm c\sqrt{k^2 + (m_0 c \setminus \hbar)^2}$, where $\hbar = \frac{h}{2\pi}$ is the Dirac's constant, for the Planck's constant $h = 6.6210^{-34} Js$.

14

Note that we assume that ω can be positive or negative (clockwise or counterclockwise angular frequency), so that the energy of the particle is $E = \hbar|\omega|$, where $|_-|$ denotes the absolute value. In the rest of this manuscript we will consider the cases when ω is positive.

Notice that if we represent an elementary particle, with energy $E = \hbar\omega$ and momentum $\overrightarrow{\mathbf{p}} = \hbar\overrightarrow{\mathbf{k}}$, by a single harmonic $Ae^{-i\overrightarrow{\mathbf{k_4}}\overrightarrow{\mathbf{r_4}}}$ in this four-dimensional space, where $\overrightarrow{\mathbf{k_4}} = \frac{\omega}{c}e_0 + \overrightarrow{\mathbf{k}} = \frac{E}{\hbar c}e_0 + \frac{\overrightarrow{\mathbf{p}}}{\hbar}$, then we obtain that $k_4^2 = (\frac{\omega}{c})^2 - k^2 = 0$ for particles with rest mass equal to zero (photons, gravitons, etc..), and $k_4^2 > 0$ for the massive particles (with rest mass m_0 greater than zero). In fact, we obtain that $|k_4| = \frac{\omega_0}{c}$ where $\omega_0 = \frac{m_0 c^2}{\hbar}$ is the invariant angular frequency for particles (analog to the invariant rest mass m_0 of particles). Thus, similarly to the 3-dimensional angular wavenumber vector $\overrightarrow{\mathbf{k}}$ that in physics means the particle's momentum, the 4-dimensional angular wavenumber $\overrightarrow{\mathbf{k_4}}$ has a physical meaning as the particle's relativistically invariant angular frequency.

□

Consequently, in any given instance of time t, any matter-event in this time-space is a particular time-space perturbation $\Psi(\overrightarrow{\mathbf{r_4}})$, that can be mathematically given by the following Fourier transformation [20]:

$\Psi(\overrightarrow{\mathbf{r_4}}) = \Psi(x, y, z, t) = \int C(k_4)e^{i(-\overrightarrow{\mathbf{k_4}}\overrightarrow{\mathbf{r_4}})}dk_4 =$

$= \int A(k)e^{i(-\overrightarrow{\mathbf{k}}\overrightarrow{\mathbf{r}} - \omega(k)t)}dk =$ (where $A(k) = C(k_4) = C(\sqrt{(\frac{\omega(k)}{c})^2 - k^2})$))

$= \int \int \int_{-\infty}^{+\infty} A(k)e^{i(k_x x + k_y y + k_z z - \omega(k)t)}dk_x dk_y dk_z.$

It is a space-distribution of a particle in a given instance of time t, and it changes in time, that is, the amplitudes $A(k)$ are generally dependent on time as well.

Mathematically, these matter-events are complex functions, composed by one real and one imaginary component. The amplitudes $A(k)$ of the harmonics, in a given instance of time t, are given by inverse Fourier transformation,

$A(k) = \int \int \int_{-\infty}^{+\infty} \Psi(x, y, z, t)e^{-i(k_x x + k_y y + k_z z - \omega(k)t)}dx dx dz.$

The elementary particles are packet waves that propagate in this four-dimensional space.

Thus, for such particular *stationary* cases we have that $d\omega(k)/d\overrightarrow{\mathbf{k}}$ is constant (that is, it does not depend on $\overrightarrow{\mathbf{k}}$), equal to the particle's velocity $-\overrightarrow{\mathbf{v}} = -v_x e_1 - v_y e_2 - v_z e_3$ (negative sign is the consequence that $e_i, i \geq 1$ are imaginary, thus the scalar products of (only) spatial vectors are negative), that can depend on the time t as well.

Consequently, for any fixed instance of time t, by integration we obtain that,

(0) $\int_{\overrightarrow{\mathbf{k_0}}}^{\overrightarrow{\mathbf{k}}} d\omega = \omega(k) - \omega(k_0) = -\int_{\overrightarrow{\mathbf{k_0}}}^{\overrightarrow{\mathbf{k}}} \overrightarrow{\mathbf{v}} d\overrightarrow{\mathbf{k}} = -\overrightarrow{\mathbf{v}} \int_{\overrightarrow{\mathbf{k_0}}}^{\overrightarrow{\mathbf{k}}} d\overrightarrow{\mathbf{k}} = -\overrightarrow{\mathbf{v}}(\overrightarrow{\mathbf{k}} - \overrightarrow{\mathbf{k_0}}),$

where the constant $\overrightarrow{\mathbf{k_0}} = \frac{\overrightarrow{\mathbf{p}}}{\hbar}$ for a given momentum $\overrightarrow{\mathbf{p}} = p_x e_1 + p_y e_2 + p_z e_3$ of a particle that is collinear with the velocity $\overrightarrow{\mathbf{v}}$, that is, $\overrightarrow{\mathbf{p}}\overrightarrow{\mathbf{v}} = -pv$. Because of that we can write $\overrightarrow{\mathbf{p}} = p\overrightarrow{i_v}$, $\overrightarrow{\mathbf{v}} = v\overrightarrow{i_v}$, where $\overrightarrow{i_v}$ is unitary vector tangent to the trajectory of a particle (i.e., $\overrightarrow{i_v}\,\overrightarrow{i_v} = -1$). Thus,

$\omega(k) = \omega_0 + v_x(k_x - \frac{p_x}{\hbar}) + v_y(k_y - \frac{p_y}{\hbar}) + v_z(k_z - \frac{p_z}{\hbar}),$

where ω_0 denotes the constant $\omega(k_0)$ that does not depend on k but may depend on time as we will see in what follows.

The *phase velocity* of a particle's wave-packet, observed in a given referential system, is defined by $\vartheta = \frac{\omega_0}{k_0}$.

The constant ω_0 is determined as follows in the following two cases, by considering that the angular frequency $\omega(k)$ for its particular values is correlated by De Broglie to the total energy o particle $E = \hbar\omega(k)$:

- Case for *massive particles* (with rest mass $m_0 > 0$), denominated as mass-particles as well: when $\overrightarrow{\mathbf{v}} = 0$ then the energy of this particle is $E = m_0 c^2$, that is, the energy in the rest-state of this particle. Consequently, from (0) we have that $\omega_0 = \omega(k) = \frac{m_0 c^2}{\hbar}$.

  Consequently, for the total energy of these mass-particles that propagates with velocity $v = |\overrightarrow{\mathbf{v}}|$, with $\beta = \frac{v}{c}$, it holds that,

  $$E = \sqrt{(m_0 c^2) + (pc)^2} = \sqrt{(\hbar\omega_0)^2 + (pc)^2} = \frac{m_0 c^2}{\sqrt{1-\beta^2}} = \hbar\omega_v,$$

  where $\omega_v = \omega_0/\sqrt{1-\beta^2}$ is a computed angular frequency relative to the velocity v of this particle w.r.t. the reference system of an observer (for different observes in different referential systems, that move with different velocities, this computed value of the *same* observed particle is different). For an observer in a given fixed position (the origin of its coordinate system, for example), this observed particle's frequency is by Lorentz low slowed down by the factor $\sqrt{1-\beta^2}$, so that the *really observed* particle's angular frequency of the observed wave-packet $\Psi(\overrightarrow{\mathbf{r_4}})$ given above is constant and equal to $\omega_v\sqrt{1-\beta^2} = \omega_0$. Thus, ω_0 is the angular frequency of this massive particle equal in any inertial system (without acceleration), that is, an invariant as is the rest-mass m_0.

- Case for *massless particles* (with rest mass $m_0 = 0$): they propagate, as usual, with very high velocity $c \geq v > 0$ equal to the maximal velocity of light if this particle propagates in the vacuum, thus the zero value of equation (0) we can obtain when $\overrightarrow{\mathbf{k}} = \overrightarrow{\mathbf{k}}_0 = \frac{\overrightarrow{\mathbf{p}}}{\hbar}$. Consequently, the value of E is the *total* energy of this particle with the given momentum $\overrightarrow{\mathbf{p}}$, so that $\omega_0 = \omega(k) = \omega(\frac{p}{\hbar}) = \frac{E}{\hbar}$. The total energy of massless particles is defined by $E = \hbar\omega_0 = (\hbar k_0)\vartheta = p\vartheta$. When a particle propagates in the vacuum then $\vartheta = c$, so that $E = pc$.

  When the total energy changes in time, to a fixed observer this angular frequency appears to change as well (for example, the relativistic effects for red-shifting of photons for a fixed observer). Thus, differently from massive particles where for a fixed observer $\frac{\partial\omega_0}{\partial t} = 0$, here ω_0 can change in time, if a particle changes its total energy during the propagation.

Consequently, for the wave-packet of an elementary particle, and given reference system, we have that

$$(1) \quad \Psi(x,y,z,t) = \int A(k)e^{i(-\overrightarrow{\mathbf{k}}\overrightarrow{\mathbf{r}}-\omega(k)t)}dk =$$
$$= (\int A(k)e^{-i(\overrightarrow{\mathbf{k}}\overrightarrow{\mathbf{r}}-\overrightarrow{\mathbf{v}}(\overrightarrow{\mathbf{k}}-\overrightarrow{\mathbf{k_0}})t)}dk)e^{-i\omega_0 t} =$$
$$= (\int A(k)e^{-i(\overrightarrow{\mathbf{k}}-\overrightarrow{\mathbf{k_0}})(\overrightarrow{\mathbf{r}}-\overrightarrow{\mathbf{v}}t)}dk)e^{i(-\overrightarrow{\mathbf{k_0}}\overrightarrow{\mathbf{r}}-\omega_0 t)} =$$
$$= \Phi(\overrightarrow{\mathbf{r}},t)e^{i(-\frac{\overrightarrow{\mathbf{p}}\cdot\overrightarrow{\mathbf{r}}}{\hbar}-\omega_0 t)} = \Phi(\overrightarrow{\mathbf{r}},t)e^{i\frac{p}{\hbar}(-i_v\overrightarrow{\mathbf{r}}-\vartheta t)}.$$

In what follows we introduce the spatial vector $\overrightarrow{\mathbf{u}} = \overrightarrow{\mathbf{r}} - \overrightarrow{\mathbf{v}}t$, that is equal to zero for the time-space points of particle's trajectory.

16

The "corpuscular" geometric wave-packet shape (matter's distribution) of a particle, that *appears to a fixed observer*, is given by

$\Phi(\overrightarrow{\mathbf{r}}, t) = \Phi(x, y, z, t) = \int A(k) e^{-i(\overrightarrow{\mathbf{k}} - \overrightarrow{\mathbf{k_0}})(\overrightarrow{\mathbf{r}} - \overrightarrow{\mathbf{v}} t)} dk =$

$= \int A(|\overrightarrow{\mathbf{k}} + \overrightarrow{\mathbf{k_0}}|) e^{-i \overrightarrow{\mathbf{k}} \overrightarrow{\mathbf{u}}} dk =$

(here we denote by $B(k)$ the value $A(|\overrightarrow{\mathbf{k}} + \overrightarrow{\mathbf{k_0}}|)$),

$= \int\int\int_{-\infty}^{+\infty} B(k) e^{i(k_x(x - v_x t) + k_y(y - v_y t) + k_z(z - v_z t))} dk_x dk_y dk_z =$

$= \int\int\int_{-\infty}^{+\infty} B(k) e^{i(k_x u_x + k_y u_y + k_z u_z)} dk_x dk_y dk_z.$

In the case when a particle propagates in the vacuum with a constant velocity $\overrightarrow{\mathbf{v}}$ (stationary case), then the coefficients $B(k)$ does not change in time, i.e. $\frac{\partial B(k)}{\partial t} = 0$, so that the "corpuscular" geometry (matter distribution) does not change in time and $\Phi(\overrightarrow{\mathbf{r}}, t) = \Phi(\overrightarrow{\mathbf{u}}) = \Phi(\overrightarrow{\mathbf{r}} - \overrightarrow{\mathbf{v}} t) = \Phi(x - v_x t, y - v_y t, z - v_z t)$ is a wave-packet that propagates with a velocity $\overrightarrow{\mathbf{v}}$.

For instance, in such a stationary case, for a particle with $m_0 = 0$ (for example boson as photon, graviton, etc..) that propagates in the vacuum with a velocity equal to its phase velocity (when $c = |\overrightarrow{\mathbf{c}}| = \vartheta$), so that the total energy is $E = p\vartheta = pc = -\overrightarrow{\mathbf{p}} \overrightarrow{\mathbf{c}}$, we have that $\omega_0 = -\overrightarrow{\mathbf{p}} \overrightarrow{\mathbf{c}}/\hbar$, so that

$\Psi(x, y, z, t) = \Phi(\overrightarrow{\mathbf{r}} - \overrightarrow{\mathbf{c}} t) e^{-i \overrightarrow{\mathbf{p}}(\overrightarrow{\mathbf{r}} - \overrightarrow{\mathbf{c}} t)/\hbar} = \Psi(\overrightarrow{\mathbf{r}} - \overrightarrow{\mathbf{c}} t).$

But in non-stationary cases, when a particle changes its velocity and total energy, its distribution ("corpuscular" geometry) Φ changes in time as well, because the coefficients $B(k)$ in the Fourier integral above change in time, that is, $\frac{\partial B(k)}{\partial t} \neq 0$. In such more complex cases with an acceleration of particles, we have both phenomena: particle changes its velocity, momentum and total energy (because of the interaction with another particles), and its corpuscular geometry as well.

There are particular cases, when a particle does not change its total energy and with $\frac{\partial}{\partial t}(\overrightarrow{\mathbf{p}} \overrightarrow{\mathbf{v}}) = 0$, with acceleration caused only by (constant) velocity that changes its direction, when Φ does not change in time its distribution w.r.t. its trajectory, so that it is of the form $\Phi(\overrightarrow{\mathbf{r}} - \overrightarrow{\mathbf{v}} t)$. For instance in the case of the stationary rotation of electrons around nucleus of an atom.

Thus, based on standard Fourier transformation, the function $\Phi(\overrightarrow{\mathbf{r}}, t)$ is a *real* function, differently from $\Psi(x, y, z, t)$ that is *complex*. The real and imaginary components of $\Psi(x, y, z, t)$ are determined by the oscillation of the complex oscillator component $e^{i(-\frac{\overrightarrow{\mathbf{p}} \overrightarrow{\mathbf{r}}}{\hbar} - \omega_0 t)}$, that is an oscillation identical to the complex *plain wave* (like, for example, the complex electromagnetic plain wave). The amplitudes $B(k)$ of the harmonics can be obtained by the inverse Fourier transformation, for each given instance of time t, by: $B(k) = \int\int\int_{-\infty}^{+\infty} \Phi(x, y, z, t) e^{-i(k_x u_x + k_y u_y + k_z u_z)} dk_x dk_y dk_z$.

Thus, generally any particle is determined by the wave-packet $\Psi(x, y, z, t)$ composed by two sub components: by the corpuscular matter distribution $\Phi(x, y, z, t)$ that is a real function, and by the 'phase wave' $e^{i(-\frac{\overrightarrow{\mathbf{p}} \overrightarrow{\mathbf{r}}}{\hbar} - \omega_0 t)}$ that is a complex function.

Example 1: speed of light.

When light propagates through a material, it travels slower than the vacuum speed. This is a change in the phase velocity ϑ of the light and is manifested in physical effects such as refraction. The ratio between c and the speed ϑ at which light travels in a material is called the refractive index n of the material ($n = c/\vartheta$). Refraction occurs

17

when light waves travel from a medium with a given refractive index to a medium with another at an angle. At the boundary between the media, the wave's phase velocity is altered, usually causing a change in direction. Its wavelength increases or decreases but its frequency remains constant (thus photons change their momentum and velocity but does not change their energy). This reduction in speed is quantified by the refractive index of the material.

For example, for visible light the refractive index of glass is typically around 1.5, meaning that light in glass travels at $c/1.5 \approx 200,000 km/s$; the refractive index of air for visible light is about 1.0003, so the speed of light in air is very close to c.

Certain materials have an exceptionally low *group* velocity for light waves, a phenomenon called slow light. In 1999, a team of scientists led by Lene Hau were able to slow the speed of a light pulse to about 17 meters per second (61 km/h; 38 mph) [13], they were able to momentarily stop a beam [6].

□

We will show that, in the case of these massless elementary particles that propagate in vacuum with the velocity of light c, we have that this "corpuscular" wave-packet in the *stable*(general) state corresponds to the Dirac function (with $A(k) = B(k) = 1/(2\pi)^3$), $\delta(\overrightarrow{r} - \overrightarrow{c} t) = \delta(x - c_x t, y - c_y t, z - c_z t) = \delta(x - c_x t)\delta(y - c_y t)\delta(z - c_z t)$.

It is reasonable assumption that the volume of a distribution Φ (where it is greater than zero) of a massive particle (with rest mass m_0 greater than zero) is always greater than zero, so that it is a reason that such particles can not reach the limit velocity of light. In the analog way, the massless particles (with rest mass equal to zero) must have, in their stable state, this volume equal to zero, so that their distribution Φ is equal to Dirac function above, and they are able to propagate with the velocity of light in the vacuum. The non stable states of particles with rest mass $m_0 = 0$ can have more complex wave-packet forms and it happen only in a very short instances of time, when the particle enters in strongly unsymmetric space region, as will be explained in what follows. In such situations its velocity of propagation becomes less than the velocity of light in the vacuum so that this particle can have a similar behavior as massive particles with Φ that occupies a limited but nonzero volume (so called spatial explosion of excited bosons). This unstable state of the particles with $m_0 = 0$ tends to come back into the stable state with Dirac function geometry for distribution Φ.

For any matter-perturbation of an elementary particle that propagates in the 3-dimensional space with a velocity that changes in the time, because of external forces that influence this particle, the 3-dimensional wave-packet distribution $\Phi(x, y, z, t)$ changes as well, but it must satisfy the *conservation matter* principle, that is, at each given time instance t it must be satisfied the following *invariance* property:

(2) $\quad 1_\Phi = \int\int\int_{-\infty}^{+\infty} \Phi(x, y, z, t) dx dy dz = \int\int\int_{-\infty}^{+\infty} \Phi(x, y, z, 0) dx dy dz > 0,$

where 1_Φ is a time-invariant constant value of an elementary particle (not necessarily equal to 1), and $V(t)$ is a finite cube (or sphere) which contains the whole "corpuscular" wave-packet in a given instance of time t, and $dV = dx dy dz$.

We define the minimal (limit) cube $V_m(t) = lim(2^3 \triangle X \triangle Y \triangle Z)$, such that in this time-instance t, $\Phi(x, y, z, t) = 0$ for $(x \leq -\triangle X$ or $x \geq \triangle X$ or $y \leq -\triangle Y$ or $y \geq \triangle Y$ or $z \leq -\triangle Z$ or $z \geq \triangle Z)$.

The real function $\Phi(x, y, z, t)$ is the "corpuscular" geometric wave-packet form of a

18

particle that propagates in the ordinary 3-dimensional space with a velocity $\overrightarrow{\mathbf{v}} = v_x e_1 + v_y e_2 + v_z e_3$). In the stationary case, when it propagates in the vacuum with a constant velocity, it has constant distribution, that propagates as wave-packet $\Phi(\overrightarrow{\mathbf{u}}) = \Phi(\overrightarrow{\mathbf{r}} - \overrightarrow{\mathbf{v}}t) = \Phi(x - v_x t, y - v_y t, z - v_z t)$.

The interactions between any two wave-packets (particles) can be obtained only by their local collisions, and depending on their energy and velocities they can produce a kind of Compton effects (elastic collisions) where they survive the collisions by chaining their momentum and energy (with conservation of total momentum and energy), or can make total fusion between them with possible creation of new stable particles (in Feynman's diagrams). In order to be able for two wave-packets to have a collision, and mutual interference, at least one of them must have a volume $V_m(t)$ (in a given instance of time of mutual collision) greater than zero. So, from this point of view, it can not happen that the distance between any two particles becomes equal to zero, so that we avoid classic infinitary problems of gravitational and electronic fields and forces where the particles are pointlike, so that it is possible to have the distances between particles equal to zero with, consequently, infinite values of gravitational (or electric) forces.

The particles with $V_m(t)$ equal to zero are, for example, the particles with $\Phi(x - v_x t, y - v_y t, z - v_z t)$ equal to the Dirac function $\delta(x - v_x t, y - v_y t, z - v_z t) = \delta(x - v_x t)\delta(y - v_y t)\delta(z - v_z t)$.

Thus, for any two particles with the "corpuscular" form given by Dirac function, it is impossible to have the collisions in their stable states, but only when they are excited and are involved in their temporary "spatial explosions". Such explosions can happen also when two stable particles are very close one to another so that the ideal spatial symmetry for a free particle in the vacuum does not hold more: it explains why, for example, photons can interact with gravitons (i.e., gravitational field) and may have the gravitational redshifts. Because of that, it will be natural consequence that the massless particles, as bosons (gravitons, photons, etc..) in their stable states, will have the volume $V_m(t)$ equal to zero (with Dirac function for their distribution Φ). In that case they can be used as intermediators between the massive particles (that have the rest mass and the volume $V_m(t)$ greater than zero, that is, to be the quantum-correspondence for the "fields" (the statistical events as gravitational, electromagnetic, etc., that are statistical results of actions of a high number of bosons), by avoiding in more common situations the significant interference between themselves.

This situation can be obtained in the quantum level only if the collisions between gravitons and photons, for example, are practically improbable. Consequently, a number of gravitons and photons can coexist in the same small region of space without any significant direct interference between them during the contemporary collisions with particles with rest mass and volume $V_m(t)$ greater than zero. Also in such situation we can have the cases of the interference between a graviton and a photon, in the situation when they are very, very close in a given instance of time, so that they are involved in temporary spatial explosion (with their volume of distribution greater than zero in that instance of time). In normal situations, these interferences statistically can be neglected, while in the cases of very strong field interactions (when the local density of photons and gravitons is very high) these inter-boson's interactions are significant, so that the gravitation field has strong interactions with the electromagnetic field in a given local space region.

19

2.2 Energy, mass and momentum of the time-space perturbations

Analogously to the Schrodinger's approach, used to derive its equation based on total energies of particles, with the mapping $E \rightarrow i\hbar\frac{\partial}{\partial t}$, that does not take in consideration the spatial matter-distribution of a particle (considering in standard quantum theory only the pointlike particles), here, differently, we define the total energy E in a given time instance t, of the time-space perturbations defined by wave-packets (1) and (2), by taking in consideration its real spatial matter distribution. Thus, by definition of a *spatial integral* as follows:

(3) $\quad E = |\int_{-\infty}^{+\infty} i\hbar \frac{\partial \Phi(\overrightarrow{r},t)e^{-i\omega_0 t}}{\partial t} dV|/1_{\Phi} = |\oint_{V_m(t)} i\hbar \frac{\partial \Phi(\overrightarrow{r},t)e^{-i\omega_0 t}}{\partial t} dV|/1_{\Phi}$,

where $dV = dxdydz$.

Let $\nabla = e_1 \frac{\partial}{\partial x} + e_2 \frac{\partial}{\partial y} + e_3 \frac{\partial}{\partial z}$ be the gradient, so that the Laplacian is defined by $\triangle = -\nabla^2 = \frac{\partial^2}{\partial x^2} + \frac{\partial^2}{\partial y^2} + \frac{\partial^2}{\partial z^2}$. Then the derivation of the wave-packet along its trajectory, with the unitary tangent vector $\overrightarrow{i_v}$ on the trajectory, collinear with the vector of its velocity $\overrightarrow{v} = v\overrightarrow{i_v} = \overrightarrow{i_v}\sqrt{v_x^2 + v_y^2 + v_z^2}$, is denoted by the operator $\overrightarrow{i_v}\nabla$.

In what follows, by $Re()$ and $Im()$ we will denote the real and imaginary component of complex expressions, so that $\gamma = Re(\gamma) + iIm(\gamma)$ for a complex number γ.

Proposition 1 *The geometric distribution $\Phi(\overrightarrow{r},t)$ of the wave-packets with the independent space-time variables $\overrightarrow{r}$ and t, of an elementary particle given in (1), is determined by the following differential equations:*

(e.1) $\quad e^{i\omega_0 t}\frac{\partial \Phi e^{-i\omega_0 t}}{\partial t} = -i\frac{\partial(\omega_0 t)}{\partial t})\Phi + \overrightarrow{v_1}\nabla\Phi + \Phi_D(\overrightarrow{r},t)$, *where,*

$\overrightarrow{v_1} = \frac{\partial}{\partial t}(\overrightarrow{v}t)$, $\Phi_D(\overrightarrow{r},t) = \int_{-\infty}^{+\infty} \frac{\partial B(k)}{\partial t}e^{i(k_x(x-v_xt)+k_y(y-v_yt)+k_z(z-v_zt))}dk_xdk_ydk_z$

is equal to zero when this particle is in a stationary state, that is, when $\frac{\partial B(k)}{\partial t} = 0$. Thus,

(e.2) $\quad \frac{\partial \Psi}{\partial t} = -i\omega_p \Psi + \overrightarrow{v_1}\nabla\Psi + \Psi_D(\overrightarrow{r},t)$,

where $\Psi_D(\overrightarrow{r},t) = \Phi_D(\overrightarrow{r},t)e^{i(-\frac{\overrightarrow{p}\cdot\overrightarrow{r}}{\hbar}-\omega_0 t)}$, with $\omega_p = \omega_1 + \frac{\overrightarrow{p}\cdot\overrightarrow{v_1}}{\hbar}$ and $\omega_1 = \frac{\partial}{\partial t}(\frac{\overrightarrow{p}\cdot\overrightarrow{r}}{\hbar} + \omega_0 t) = \frac{\overrightarrow{r}}{\hbar}\frac{\partial\overrightarrow{p}}{\partial t} + \frac{\partial}{\partial t}(\omega_0 t)$, where the particle's velocity $\overrightarrow{v}$, momentum $\overrightarrow{p}$, and ω_0 in the case of massless particles, may change in time.

Proof: From the fact (1) we have that
$\frac{\partial}{\partial t}(\Phi e^{-i\omega_0 t}) = -i\frac{\partial(\omega_0 t)}{\partial t}\Phi e^{-i\omega_0 t} + (\frac{\partial\Phi}{\partial t})e^{-i\omega_0 t}$.

Let us show that $\frac{\partial\Phi}{\partial t} = \frac{\partial(\overrightarrow{v}t)}{\partial t}\nabla\Phi + \Phi_D$. We can make the following derivation:

(a.0) $\quad \frac{\partial\Phi}{\partial t} = \frac{\partial}{\partial t}(\int\int\int_{-\infty}^{+\infty} B(k)e^{i(k_x(x-v_xt)+k_y(y-v_yt)+k_z(z-v_zt))}dk_xdk_ydk_z)$

$= \int_{-\infty}^{+\infty}(-B(k)i(\frac{\partial(v_xt)}{\partial t}k_x + \frac{\partial(v_yt)}{\partial t}k_y + \frac{\partial(v_zt)}{\partial t}k_z) + \frac{\partial B(k)}{\partial t})e^{i(k_x(x-v_xt)+k_y(y-v_yt)+k_z(z-v_zt))}dk$

$= -\frac{\partial(v_xt)}{\partial t}\int\int\int_{-\infty}^{+\infty}(ik_x)B(k)e^{i(k_x(x-v_xt)+k_y(y-v_yt)+k_z(z-v_zt))}dk_xdk_ydk_z - \ldots$

$- \frac{\partial(v_zt)}{\partial t}\int\int\int_{-\infty}^{+\infty}(ik_z)B(k)e^{i(k_x(x-v_xt)+k_y(y-v_yt)+k_z(z-v_zt))}dk_xdk_ydk_z + \Phi_D$

$= -\frac{\partial(v_xt)}{\partial t}\frac{\partial\Phi}{\partial x} - \frac{\partial(v_yt)}{\partial t}\frac{\partial\Phi}{\partial y} - \frac{\partial(v_zt)}{\partial t}\frac{\partial\Phi}{\partial z} + \Phi_D = \frac{\partial(\overrightarrow{v}t)}{\partial t}\nabla\Phi + \Phi_D$.

Consequently, we obtain,

(a.1) $\quad \frac{\partial}{\partial t}(\Phi e^{-i\omega_0 t}) = (-i\omega_0\Phi - it\frac{\partial\omega_0}{\partial t}\Phi + \frac{\partial(\overrightarrow{v}t)}{\partial t}\nabla\Phi + \Phi_D)e^{-i\omega_0 t}$, and the differential equation (e.1).

From the fact that,

(a.2) $\nabla(\Phi e^{-i\frac{\vec{p}\cdot\vec{r}}{\hbar}}) = (\nabla\Phi)e^{-i\frac{\vec{p}\cdot\vec{r}}{\hbar}} + \Phi\frac{i}{\hbar}(p_x e_1 + p_y e_2 + p_z e_3)e^{-i\frac{\vec{p}\cdot\vec{r}}{\hbar}} =$

$= (\nabla\Phi + \frac{i}{\hbar}\vec{p}\Phi)e^{-i\frac{\vec{p}\cdot\vec{r}}{\hbar}},$

we obtain that $(\nabla\Phi)e^{-i\frac{\vec{p}\cdot\vec{r}}{\hbar}} = \nabla(\Phi e^{-i\frac{\vec{p}\cdot\vec{r}}{\hbar}}) - \frac{i}{\hbar}\vec{p}\Phi e^{-i\frac{\vec{p}\cdot\vec{r}}{\hbar}}$, and

(a.3) $(\nabla\Phi)e^{i(-\frac{\vec{p}\cdot\vec{r}}{\hbar} - \omega_0 t)} = \nabla\Psi - \frac{i}{\hbar}\vec{p}\Psi,$

and by substitution in (e.1) we obtain (e.2).

□

Notice that $\frac{\partial\omega_0}{\partial t} \neq 0$ only for the massless particles in their unstable states and very-very short interval of times $\triangle t \approx 0$, when they change their total energy $E = \hbar\omega_0$ during the collisions with another particles (the Compton effects). In what follows we will denote by $\mathcal{E} = \hbar\omega_p$ the energy associated to this angular frequency ω_p.

Example 2: The *stationary case* is obtained when a particle propagates with constant total energy E and constant value $\mathcal{E} = \hbar\omega_p$. Thus, in such a stationary case we have that $\Phi_D(\vec{r}, t) = 0, \Psi_D(\vec{r}, t) = 0$. It is easy to verify that the stationary case is one, for example, of the following two cases:

1. When a particle propagates with constant momentum $\vec{p}$, velocity $\vec{v}$ (thus, $\frac{\partial\vec{v}}{\partial t} = \frac{\partial\vec{p}}{\partial t} = 0$), and total energy E (in that case ω_0 is constant for massless particles as well). Thus, without any acceleration. In that case ω_p is constant as well, with constant $\mathcal{E} = \hbar(\frac{\vec{r}}{\hbar}\frac{\partial\vec{p}}{\partial t} + \frac{\partial}{\partial t}(\omega_0 t) + \frac{\vec{p}\vec{v}}{\hbar}) = \hbar\omega_0 + \vec{p}\vec{v} = \hbar\omega_0 - pv$.

2. When a particle routes with constant radius R around a fixed center, with constant angular velocity $\nu = \frac{|\vec{v}|}{R} = \frac{v}{R}$, constant value of the momentum $p = |\vec{v}|$, and total energy E. In this case we can obtain ω_p constant in a particular co-ordinate system: coordinate center of the reference system x, y of the plain in which this particles routes. Then, the position of the trajectory of this particle in a given moment t is equal to $\vec{r} = R\vec{i_\theta}$, where $\vec{i_\theta}$ is a unitary radial vector with angle $\theta = \nu t$ w.r.t the axis x. In this case the acceleration $\frac{\partial\vec{v}}{\partial t} = -|\frac{\partial\vec{v}}{\partial t}|\vec{i_\theta}$, and $\frac{\partial\vec{p}}{\partial t} = -|\frac{\partial\vec{p}}{\partial t}|\vec{i_\theta}$ are radial vectors that have the constant values and are orthogonal to the vectors of velocity $\vec{v}$ and momentum $\vec{p}$. So that, $\vec{p}\vec{r} = 0$, $\vec{p}\frac{\partial\vec{v}}{\partial t} = 0$, and
$\omega_p = \frac{\partial}{\partial t}(\frac{\vec{p}\vec{r}}{\hbar} + \omega_0 t) + \frac{\vec{p}\vec{v}}{\hbar} = +\omega_0 + \frac{1}{\hbar}\vec{p}(\vec{v} + t\frac{\partial\vec{v}}{\partial t}) = \omega_0 + \frac{1}{\hbar}\vec{p}\vec{v} =$
$= \omega_0 - \frac{pv}{\hbar}$.
Here both ω_0 and pv are constant, and, consequently, we obtained the stationarity condition $\frac{\partial\omega_p}{\partial t} = 0$ in all points of the trajectory of this particle, with constant $\mathcal{E} = \hbar\omega_0 + \vec{p}\vec{v} = \hbar\omega_0 - pv$. This case will be applied for the stationary orbits of electrons in an atom, in Example 6.

Notice that in both stationary cases above we obtained that this particular constant energy is equal to $\mathcal{E} = \hbar\omega_p = \hbar\omega_0 + \vec{p}\vec{v} = \hbar\omega_0 - pv$.

If $\omega_p = \omega_1 + \frac{\vec{p}\vec{v}}{\hbar}$ is computed for current space-time positions on the particle's trajectory, then, in this particular case, ω_1 is taken as a derivation $\frac{\partial}{\partial t}$ of the current particle's phase $\varphi = -\frac{\vec{p}\vec{r}}{\hbar} - \omega_0 t$ (we have that $\Psi(\vec{r}, t) = \Phi(\vec{r}, t)e^{i\varphi}$) and express the parti-

21

cle's phase-changing on its trajectory. □

Notice that the energy changes only during collisions with another particles (Compton effects; we consider a "field" as a statistical result of the iterations with bosons of this particular field), so that after it this particle continue to propagate again as a stationary particle, but with new values of total energy E, velocity $\overrightarrow{\mathbf{v}}$, momentum $\overrightarrow{\mathbf{p}}$, and new stable wave-packet geometry (distribution) $\Phi(\overrightarrow{\mathbf{r}} - \overrightarrow{\mathbf{v}}t)$, so that it can be described by the simpler stationary-case (defined in Example 2) differential equations:

(e.1.1) $\quad e^{i\omega_0 t}\frac{\partial \Phi e^{-i\omega_0 t}}{\partial t} = (-i\omega_0 + \frac{\partial(\overrightarrow{\mathbf{v}}t)}{\partial t}\nabla)\Phi$

(e.2.1) $\quad \frac{\partial \Psi}{\partial t} = -\frac{i}{\hbar}\mathcal{E}\Psi + \frac{\partial(\overrightarrow{\mathbf{v}}t)}{\partial t}\nabla\Psi,$

where, when a velocity $\overrightarrow{\mathbf{v}}$ is constant, we have that $\frac{\partial(\overrightarrow{\mathbf{v}}t)}{\partial t} = \overrightarrow{\mathbf{v}} + t\frac{\partial \overrightarrow{\mathbf{v}}}{\partial t} = \overrightarrow{\mathbf{v}}$. Notice that, as in Example 2 above, we have particular stationary cases when $\overrightarrow{\mathbf{v}}$ is not constant, as for instance, for a stationary electron that rotates around the nucleus of an atom with a constant radial acceleration (in that case $\overrightarrow{\mathbf{p}}\,\overrightarrow{\mathbf{v}}$ and total energy of this electron are constant, thus $\mathcal{E} = E + \overrightarrow{\mathbf{p}}\,\overrightarrow{\mathbf{v}}$ is constant as well).

Thus, the more general equations (e.1) and (e.2) can be used only for an infinitesimal interval of time $\triangle t \approx 0$ during inter-collisions of two particles(Compton effects), when they rapidly change their energy, velocity and momentum (established by conservation of the total momentum and energy during such collisions). Thus, in the next analysis of their properties we will consider only their stationary states:

Proposition 2 *The geometric distribution of a stationary wave-packet $\Phi(\overrightarrow{\mathbf{r}} - \overrightarrow{\mathbf{v}}t)$ that propagates with a constant velocity $\overrightarrow{\mathbf{v}}$, momentum $\overrightarrow{\mathbf{p}}$, and total energy E, must satisfy the following properties:*

(4) $\quad \omega_0 = -\frac{\hbar}{\mathbf{1}_\Phi}\oint_{V_m(t)} Im(e^{i\omega_0 t}\frac{\partial}{\partial t}(\Phi e^{-i\omega_0 t}))dV.$

The rest and relativistic mass of the wave-packet of a massive particle $(m_0 > 0)$ are:

(5) $\quad m_0 = \frac{\hbar}{c^2 \mathbf{1}_\Phi}(\cos(\omega_0 t)\oint_{V_m(t)}\frac{\partial\Phi \sin(\omega_0 t)}{\partial t}dV - \sin(\omega_0 t)\oint_{V_m(t)}\frac{\partial\Phi \cos(\omega_0 t)}{\partial t}dV);$

(6) $\quad m = \frac{\hbar}{c\mathbf{1}_\Phi}\oint_{V_m(t)}(\overrightarrow{i_v}\nabla)\Phi(x - v_xt, y - v_yt, z - v_zt)dV.$

For the case when Φ is the Dirac function we obtain from (5) that $m_0 = 0$, and this is the case of stable-state particles with rest mass equal to zero, when:

(7) $\quad \Psi(x, y, z, t) = \delta(\overrightarrow{\mathbf{r}} - \overrightarrow{\mathbf{v}}t)e^{i(-\overrightarrow{\mathbf{p}}\,\overrightarrow{\mathbf{r}} - Et)/\hbar},$

where $\delta(\overrightarrow{\mathbf{r}} - \overrightarrow{\mathbf{v}}t) = \delta(x - v_xt, y - v_yt, z - v_zt)$ is the Dirac function, that propagates with a velocity $\overrightarrow{\mathbf{v}}$, equal to the velocity of light when it propagates in the vacuum. In the case when $\mathcal{E} = 0$, then $E = pv$, thus $v = c$, and we obtain the particular solution $\Psi(\overrightarrow{\mathbf{r}} - \overrightarrow{\mathbf{c}}t) = \delta(\overrightarrow{\mathbf{r}} - \overrightarrow{\mathbf{c}}t)e^{-i\overrightarrow{\mathbf{p}}(\overrightarrow{\mathbf{r}} - \overrightarrow{\mathbf{c}}t)/\hbar}.$

Proof: In this stationary case we have that $\frac{\partial(\overrightarrow{\mathbf{v}}t)}{\partial t} = \overrightarrow{\mathbf{v}} + t\frac{\partial\overrightarrow{\mathbf{v}}}{\partial t} = \overrightarrow{\mathbf{v}}$ and $\frac{\partial\omega_0}{\partial t} = 0$. From (a.1), in the proof of Proposition 1, we obtain that $\omega_0\Phi = -Im(e^{i\omega_0 t}\frac{\partial}{\partial t}(\Phi e^{-i\omega_0 t}))$ (and we obtain the result in (4), that is, $\omega_0\Phi = \cos(\omega_0 t)\frac{\partial\Phi \sin(\omega_0 t)}{\partial t} - \sin(\omega_0 t)\frac{\partial\Phi \cos(\omega_0 t)}{\partial t}.$

Thus, by integration of both sides, from the fact that $\int \Phi dV = \mathbf{1}_\Phi$ and that for particles with rest mass greater than zero $\omega_0 = m_0c^2/\hbar$, we obtain the expression (5) for the rest mass of the particles (wave-packets).

In the case when Φ is the Dirac function, and taking the axis x as a direction of propagation, we obtain the reduction:

$m_0\mathbf{1}_\Phi \qquad\qquad = \qquad\qquad \frac{\hbar}{c^2}(\cos(\omega_0 t)\oint_{V_m(t)}\frac{\partial\delta(x-vt)\sin(\omega_0 t)}{\partial t}dx \qquad -$

22

$\sin(\omega_0 t) \oint_{V_m(t)} \frac{\partial \delta(x-vt)\cos(\omega_0 t)}{\partial t} dx) =$

$= \frac{\hbar}{c^2}(\cos(\omega_0 t) \oint_{V_m(t)} \frac{\partial}{\partial t}\delta(x - vt)\Sigma_{n=0,1,...}A_{0n}(\omega_0(t - t_1))^n dx -$
$\sin(\omega_0 t) \oint_{V_m(t)} \frac{\partial}{\partial t}\delta(x - vt)\Sigma_{n=0,1,...}B_{0n}(\omega_0(t - t_1))^n dx)$, where $t_1 = \frac{x}{v}$ and
$A_{0n}, B_{0n}, n = 0, 1, 2, ..$ are the coefficients of the Taylor series of the functions
$\sin(\omega_0 t)$ and $\cos(\omega_0 t)$ near the point t_1.

This integration is significant only in the interval $vt_1 + \varepsilon \leq x = vt \leq vt_1 + \varepsilon$ for
$\varepsilon \mapsto 0$. So, we obtain

$m_0 \mathbf{1}_\Phi = \frac{\hbar}{c^2} Lim_{t \mapsto t_1}(\cos(\omega_0 t)\Sigma_{n=0,1,...}A_{0n}(\omega_0(t - t_1))^n -$
$\sin(\omega_0 t)\Sigma_{n=0,1,...}B_{0n}(\omega_0(t - t_1))^n) \oint_{V_m(t)} \frac{\partial \delta(x-vt)}{\partial t} dx = 0,$

because for the *unit doublet* we have that $\oint_{V_m(t)} \frac{\partial \delta(x-vt)}{\partial t} dx = 0$. That is, for the
geometric distribution of the wave-packet equal to the Dirac function (thus, with the
volume $V_m(t)$ of such a particle *equal to zero*), we obtain that the rest mass is equal to
zero. Consequently, the particles with the rest mass greater than zero must occupy a
finite geometric space that is greater than zero.

For these particles with $m_0 = 0$ that propagate in the vacuum with the velocity of light
c, and for $\mathcal{E} = 0$, that is with total energy $\hbar\omega_0 = E = -\overrightarrow{\mathbf{p}} \overrightarrow{\mathbf{c}} = pc$, from (1) we have
that:

$\Psi(x, y, z, t) = \delta(\overrightarrow{\mathbf{r}} - \overrightarrow{\mathbf{c}}t)e^{i(-\frac{\overrightarrow{\mathbf{p}} \cdot \overrightarrow{\mathbf{r}}}{\hbar} - \omega_0 t)} = \delta(\overrightarrow{\mathbf{r}} - \overrightarrow{\mathbf{c}}t)e^{-i\overrightarrow{\mathbf{p}}(\overrightarrow{\mathbf{r}} - \overrightarrow{\mathbf{c}}t)/\hbar} =$
$= \Psi(\overrightarrow{\mathbf{r}} - \overrightarrow{\mathbf{c}}t),$

that is, we obtain the expression (7). It is easy to verify that this wave-packet is the
solution of the differential equation (e.2.1) for the case when $\mathcal{E} = 0$, that is, when the
phase velocity $\vartheta = \frac{\hbar\omega_0}{p}$ is equal to the particle's velocity v.

The relativistic mass of the wave-packet with $m_0 > 0$ in a given time instance t can be
derived as follows: from (e.1.1) we have that,

$e^{i\omega_0 t} \int \int \int_{-\infty}^{+\infty} i\hbar\frac{\partial}{\partial t}(\Phi e^{-i\omega_0 t})dxdydz =$
$= \hbar\omega_0 \int \int \int_{-\infty}^{+\infty} \Phi dxdydz + i\hbar v \int \int \int_{-\infty}^{+\infty}(\overrightarrow{i_v}\nabla)\Phi(x, y, z, t)dxdydz =$
$= \hbar\omega_0 \mathbf{1}_\Phi + i\hbar v \int \int \int_{-\infty}^{+\infty}(\overrightarrow{i_v}\nabla)\Phi dxdydz$, because from the invariance property (2)
of the space-time perturbations we have that $\int \int \int_{-\infty}^{+\infty} \Phi dxdydz = \mathbf{1}_\Phi$.

Notice that $\Phi(x - v_x t, y - v_y t, z - v_z t)$ is a geometry of the wave-packet during the
propagation, and a *real* function, thus,

(a.2) $(\hbar\omega_0)^2 + (\hbar v \int \int \int_{-\infty}^{+\infty}(\overrightarrow{i_v}\nabla)\Phi dxdydz)^2/\mathbf{1}_\Phi^2 =$
$= |e^{i\omega_0 t} \int \int \int_{-\infty}^{+\infty} i\hbar\frac{\partial}{\partial t}(\Phi e^{-i\omega_0 t})dxdydz|^2/\mathbf{1}_\Phi^2 =$
$= |\int \int \int_{-\infty}^{+\infty} i\hbar\frac{\partial}{\partial t}(\Phi e^{-i\omega_0 t})dxdydz|^2/\mathbf{1}_\Phi^2 = E^2$, from (2).

But, from the relativistic theory for particles with $m_0 > 0$ we have that
$E^2 = (m_o c^2)^2 + (pc)^2 = (\hbar\omega_0)^2 + (mvc)^2$. Thus, we have that
$mvc = (\hbar v \int \int \int_{-\infty}^{+\infty}(\overrightarrow{i_v}\nabla)\Phi dxdydz)/\mathbf{1}_\Phi$, that is, we obtain (6):

$m = \frac{\hbar}{c\mathbf{1}_\Phi} \int \int \int_{-\infty}^{+\infty}(\overrightarrow{i_v}\nabla)\Phi(x - v_x t, y - v_y t, z - v_z t)dxdydz$
$= \frac{\hbar}{c\mathbf{1}_\Phi} \oint_{V_m(t)}(\overrightarrow{i_v}\nabla)\Phi(x - v_x t, y - v_y t, z - v_z t)dV.$

For the particles with $m_0 = 0$ in the stationary (stable) states, when $E = \omega_0\hbar$, from
(a.2) we obtain:

$(E)^2 + (\hbar v \int \int \int_{-\infty}^{+\infty}(\overrightarrow{i_v}\nabla)\Phi dxdydz)^2/\mathbf{1}_\Phi^2 = E^2$, thus

23

$0 = (\hbar v \int\int\int_{-\infty}^{+\infty}(\overrightarrow{i_v}\nabla)\Phi dxdydz)^2$, and we obtain that the relativistic mass m must be equal to 0. In this case we have that

$$0 = v\tfrac{\hbar}{c}\int\int\int_{-\infty}^{+\infty}(\overrightarrow{i_v}\nabla)\Phi(x - v_xt, y - v_yt, z - v_zt)dxdydz =$$

$$= \tfrac{\hbar}{c}\int\int\int_{-\infty}^{+\infty}\overrightarrow{\mathbf{v}}\nabla\Phi(x - v_xt, y - v_yt, z - v_zt)dxdydz =$$

$$= \tfrac{\hbar}{c}\int\int\int_{-\infty}^{+\infty}\tfrac{\partial\Phi(\overrightarrow{\mathbf{u}})}{\partial t}dxdydz =$$

$$= \oint_{V(t)}\tfrac{\partial\Phi(\overrightarrow{\mathbf{u}})}{\partial t}dV,$$

where $\overrightarrow{\mathbf{u}} = \overrightarrow{\mathbf{r}} - \overrightarrow{\mathbf{v}}t$.

This fact that, in each instance of time t, it holds that $\oint_{V(t)}\tfrac{\partial\Phi(\overrightarrow{\mathbf{u}})}{\partial t}dV = 0$, corresponds to the known case of the *unit doublet*: its integral over any interval enclosing zero is zero. Thus, $\Phi(\overrightarrow{\mathbf{u}}) = \delta(\overrightarrow{\mathbf{u}}) = \delta(x - v_xt)\delta(y - v_yt)\delta(z - v_zt)$ is the Dirac function, equal to 0 if $\overrightarrow{\mathbf{u}} \neq 0$, and infinite if $\overrightarrow{\mathbf{u}} \neq 0$.

These results confirm the previous considerations exposed during the computation of the rest mass for the particles with geometric wave-packet distribution equal to the Dirac function.

Consequently, from the mathematical point of view, the volume $V_m(t)$ of this particle with rest mass equal to zero in this stable (ordinary) case has to be equal to zero.

$\square$

The Dirac function for massless particles is a stationary case of propagation in the free space, but if a massless particle does not propagate in the free space but near to another particles, then we can have temporary spatial explosion with a finite volume of Φ. Dirac's function for Φ is *sufficient* condition that we will obtain the value 0 from (5), but it is not necessary condition. In fact, the massless particles ($m_0 = 0$) in a stationary state with enough big $\mathcal{E} = E - pv > \mathcal{E}_0 = E(1 - \tfrac{v}{c})$ can obtain a distribution Φ different from Dirac function, with a "materialized" rest-mass equal to:

(8) $\widehat{m_0} = \tfrac{1}{c^2}\sqrt{E^2 - (E - \mathcal{E})^2\tfrac{c^2}{v^2}} > 0.$

It explains that we can have the *massive bosons* (massive photons for example) as well. From the special relativity theory, we have that the relativistic mass is defined by $m = \tfrac{m_0}{\sqrt{1 - \tfrac{v^2}{c^2}}}$, and the total energy $E = mc^2$ with $E^2 = (m_0c^2)^2 + (pc)^2$, so that for particles with rest mass $m_0 = 0$ we have that $E = pc$ with the (nonrelativistic) mass $m = \tfrac{E}{c^2}$, invariant for any referential observer's system in the absence of acceleration. This is fundamental difference between particles with and without rest mass.

For the stationary particles with rest mass $m_0 > 0$, that propagate in the vacuum with a constant velocity $\overrightarrow{\mathbf{v}}$, we have that:

- The total energy of the wave-packet in (1), can be given by the consideration only of the real geometry (distribution) of this wave-packet:

  (9) $E = \tfrac{ch}{\mathbf{1}_\Phi}\oint_{V_m(t)}(\overrightarrow{i_v}\nabla)\Phi(x - v_xt, y - v_yt, z - v_zt)dxdydz.$

- The vector momentum, in a given time instance t, of the wave-packet for particles is defined by:

  (10) $p = |\int(-i\hbar\tfrac{\overrightarrow{\mathbf{v}}}{c}\nabla)\Phi(x - v_xt, y - v_yt, z - v_zt)dxdydz|/\mathbf{1}_\Phi.$

  It is easy to verify that the Schrödinger mapping of the momentum to the operator $(-i\hbar\nabla)$ here is dependent on the velocity of the particle and equal to operator $(-i\hbar\tfrac{\overrightarrow{\mathbf{v}}}{c}\nabla)$, i.e. proportional to the derivation of the wave-packet $\Phi(x, y, z, t)$

24

along the direction of the trajectory of this particle.

The expressions (5),(6),(9) and (10) for the rest mass, relativistic mass, total energy and momentum of a time-space perturbation of a given stationary particle (that propagates with a constant velocity, total energy and momentum), are dependent only on the geometric distribution of a wave-packet and its velocity of propagation. Consequently, as in general theory of relativity, all principal matter properties of a particle are derivable from its time-space geometry, in the way that these results can be considere as the microcosmic complements of the macrocosmic Einstein's relativistic theory as well.

It is important to notice that the wave-packet geometry $\Phi(\vec{r}, t)$ change by chaining the velocity of propagation when $\Phi_D(\vec{r}, t) \neq 0$ (when $\frac{\partial B(k)}{\partial t} \neq 0$), that is, it becomes shorter (for a fixed observer's referential system) in the direction of the propagation and its total energy, as predicted by the special theory of relativity.

But it is confirmed by expression (6) as well: in the rest-state (w.r.t. the observers referential system) and in the free homogeneous space Φ has to be symmetric in the space, that is spherical. If the velocity of this particle increments, than also the integral of the derivation of Φ along the direction of the particle's trajectory has to increment as well. The space distribution Φ will remain symmetric w.r.t. the direction of propagation but not more along this direction: the frontal side has to change its gradient more rapidly than the back side, and this is obtained by shortening its frontal part (squashed in the direction of motion), like the drop of water when propagates in the air with a big velocity. Thus, the shortening (length contraction), expected from the special theory of relativity, corresponds to a kind of shortening that increments the integral of the gradient of Φ in the direction of the propagation. The initially spherical particle during the acceleration becomes one disc (symmetric w.r.t. the axis of propagation) with frontal side that changes more rapidly than its beck side.

The compressed density of Φ in the frontal part with respect to the less compressed beck part is the reason for the lack of synchronization as viewed from another reference system for clocks separated in the direction of relative motion.

If the velocity of propagation does not change than the geometric distribution Φ is stable (stationary) and does not change with time.

This phenomena shows that inertia (inertial relativistic mass) represents the inherent dinamic property of the space. Moreover, the consequences of the results, given by Proposition 2, is that in principle *we have* the exact values of the position and of the momentum of any particle, so that Heisenberg's uncertainty principle is delimited only on the fact that we are only unable to measure precisely both of them. In fact, as we will see in the following, the "virtual particles" invented by physicists in order to talk about processes in terms of the Feynman diagrams are only unstable states of ordinary real particles, and that satisfy in each instance of time the generalized Einstein relativistic relationship between its energy $E^2 = m_0^2 c^4 + p^2 c^2$. It means that Heisenberg's uncertainty principle for the time and energy of particles is limited to the fact that we are not able to measure simultaneously the position and velocity, but they exist and are precisely defined in each instance of time. These results confirm that Einstein was right in his "debate" with Bohr (Copenhagen interpretation) and the orthodox interpretation, characterized by an acceptance of the fact that it is, as a practical matter, impossible to simultaneously determine the values of certain incompatible quantities, but the rejection that this implies that these quantities do not actually have precise values.

In fact, our results show that Einstein was right, and that probabilistic interpretation of Schrödinger equation (given by Born) is the consequence that it is only a statistical low, and as such, if we apply it to a single particle than we can only obtain the probability that statistically it will be localized in a given time-space. The "corpuscular" nature of the particles is given by their wave-packet geometry (density distribution) Φ which in each instance of time t is *localized* in a finite volume $V_m(t)$. But we will show that a stream of particles statistically results in a simple plain wave, and that each complex wave-packet Ψ of a single particle has the same diffraction, refraction, and interference effects as these statistical plain waves composed by a high number of the particles.

Now we can consider the case $m_0 = 0$ of particles with rest mass equal to zero: In their stable state, enough far from another particles, they have the Dirac function for geometric wave-packet form, $\Psi(x, y, z, t) = \delta(\overrightarrow{\mathbf{r}} - \overrightarrow{\mathbf{v}}t)\mathrm{e}^{i(-\overrightarrow{\mathbf{p}}\,\overrightarrow{\mathbf{r}}-Et)/\hbar}$, as in (7).

But there are the situations when a stable, stationary, photon becomes excited for a short interval of time, as in the situations when is sharply broken the *space symmetry* during its propagation(thus, the boundary conditions for the differential equations of particle's propagation are drastically changed): these events we will analyze later for the phenomena of refraction and "wave-behaviors" of a single photon. In all these situations a photon may change its momentum, direction of propagation and its velocity, without changing its total energy, because these "interactions" are not based on collisions with another particles (as Compton effects, or fusions), but on instantaneous space explosions of their geometric wave-packet form Φ (which is the zero-volume Dirac function id their stable states) caused by an *instantaneous* changing of the amplitudes $B(k)$ of its wave-packet harmonics $B(k)\mathrm{e}{-i\,\overrightarrow{\mathbf{k}}\,(\overrightarrow{\mathbf{r}} - \overrightarrow{\mathbf{v}}t)}$ in the presence of a local sharply broken space symmetry. These ate typical cases when we take in consideration the general equations (e.1) and (e.2) for movements of particles where the component $\Phi_D(\overrightarrow{\mathbf{r}}, t)$ is dominant, caused by the fact that in such dynamic framework we have that $\frac{\partial B(k)}{\partial t} \neq 0$ caused by a dynamical changing the boundary conditions in the local space around this particle.

For these particles with rest mass equal to zero, when $\mathcal{E} > \mathcal{E}_0$ (it can happen only when their velocity v is less than the velocity c of light in the vacuum) the situation is more complex, and for them Φ can be different from Dirac function with the volume $V_m(t)$ greater than zero, with consequence that they have an amount (8) of "materialized" rest-mass $\widehat{m_0} = \frac{1}{c^2}\sqrt{E^2 - (E - \mathcal{E})^2 \frac{c^2}{v^2}} > 0$ (the result of the inherent resistance of space to the acceleration for the particles with a volume different from zero).

Only in these particular conditions, in very short interval of times, they have an analog behavior of massive particles with volume $V_m(t)$ of its distribution Φ greater than zero (i.e., Φ in these instances of time is not more Dirac function).

It is easy to see that if the velocity v of a massless particle is $v = c$, then $\mathcal{E}_0 = 0$, so that *any* small excitation of this particle with $\mathcal{E} = E - pv > 0$ will temporarily transform this particle into a massive particle with the "materialized" rest-mass $\widehat{m_0}$ involved in a local space-explosion of its geometry distribution Φ. For smaller velocities (for example, in the case of a photon that propagates in an isotropic dielectric medium), the amount of minimal excitation $\mathcal{E}_0 > 0$ is greater in order to transform it in a temporarily massive particle.

26

Notice that this "materialized" rest-mass $\widehat{m_0}$, of the *same* particle, can change when this particle changes its velocity or total energy, differently from the massless standard particles where the rest-mass is constant and equal in every observer's referential system. Because of that, the massless particles with $m_0 = 0$ are fundamentally different from the massive particles with $m_0 > 0$, also in situations when they have a "materialized" rest-mass different from zero.

This "materialized" rest-mass $\widehat{m_0}$ of bosons can explain the following two cases in the current quantum theory, presented in the following two examples for: photons with mass and "virtual" particles.

Example 3: Current theory of a massive photon.

The photon, the particle of light which mediates the electromagnetic force is believed to be massless. The so-called Proca action describes a theory of a massive photon [12]. Classically, it is possible to have a photon which is extremely light but nonetheless has a tiny mass, like the neutrino. These photons would propagate at less than the speed of light defined by special relativity and have three directions of polarization. However, in quantum field theory, the photon mass is not consistent with gauge invariance or renormalizability and so is usually ignored. However, a quantum theory of the massive photon can be considered in the Wilsonian effective field theory approach to quantum field theory, where, depending on whether the photon mass is generated by a Higgs mechanism or is inserted in an ad hoc way in the Proca Lagrangian, the limits implied by various observations/experiments may be different [9].

An instanton is a classical solution to equations of motion with a finite, non-zero action, either in quantum mechanics or in quantum field theory. More precisely, it is a solution to the equations of motion of the classical field theory on a Euclidean spacetime. In such a theory, solutions to the equations of motion may be thought of as critical points of the action. The critical points of the action may be local maxima of the action, local minima, or saddle points. Instantons are important in quantum field theory because (a) they appear in the path integral as the leading quantum corrections to the classical behavior of a system, and (b) they can be used to study the tunneling behavior in various systems such as a Yang-Mills theory.

An instanton can be used to calculate the transition probability for a quantum mechanical particle tunneling through a potential barrier. One of the simplest examples of a system with an instanton effect is a particle in a double-well potential. In contrast to a classical particle, there is non-vanishing probability that it crosses a region of potential energy higher than its own energy.

In 3-dimensional gauge theories with Higgs fields, Hooft-Polyakov monopole's play the role of instantons. In his 1977 paper Quark Confinement and Topology of Gauge Groups, Alexander Polyakov demonstrated that instanton effects in 3-dimensional QED coupled to a scalar field lead to a *mass for the photon*.

In such a theory, its speed would depend on its frequency, and the invariant speed c of special relativity would then be the upper limit of the speed of light in vacuum [16]. The limit obtained depends on the used model: if the massive photon is described by Alexandru Proca theory [9], the experimental upper bound for its mass is about 10^{-57} grams [5]; if photon mass is generated by a Higgs mechanism, the experimental upper limit is less sharp, $m = 10^{-14} eV/c^2$ [9] (roughly $2 \cdot 10^{-47} g$).

The massive photons are particular cases of "virtual" particles. Some field interactions which may be seen in terms of virtual photons are:

- The Coulomb force (static electric force) between electric charges. It is caused by the exchange of virtual photons. In symmetric 3-dimensional space this exchange results in the inverse square law for electric force. Since the photon has no mass, the coulomb potential has an infinite range.

- The magnetic field between magnetic dipoles. It is caused by the exchange of virtual photons. In symmetric 3-dimensional space this exchange results in the inverse square law for magnetic force. Since the photon has no mass, the magnetic potential has an infinite range.

- The so-called near field of radio antennas, where the magnetic and electric effects of the changing current in the antenna wire and the charge effects of the wire's capacitive charge are detectable, but both of which effects decay with increasing distance from the antenna much more quickly than do the influence of conventional electromagnetic waves, and which are composed of real photons.

Our explanation: It is natural to suppose that in the moment of any generation of a new photon it passes from its excited unstable state with a "materialized" rest-mass and with its velocity v less than the velocity of light in the vacuum, and after very short interval of time in the free space (vacuum) of propagation it reaches its stable (stationary) state with velocity of light and geometry Φ equal to Dirac function: the distance that such a photon passes in this unstable (unstationary) initial state must be very short, and in order of dimension of one atom. In this case the interactions of electrons and photons in one atom are based on these unstable photons with "materialized" rest-mass (short distance interaction).

$\square$

Not only the photons can have in these unstable temporary states the "materialized" rest-mass, thus can be strongly influenced by gravitons (gravitational field) as massive particles, but it holds for the "virtual particles" introduced by Feynman diagrams as well:

Example 4: Current theory of Virtual particles.

In physics, a virtual particle is a particle that exists for a limited time and space, introducing uncertainty in their energy and momentum due to the uncertainty principle. Because energy and momentum in quantum mechanics are time and space derivative operators, then due to Fourier transforms their spans are inversely proportional to time duration and position spans, respectively.

Virtual particles exhibit some of the phenomena that real particles do, such as obedience to the conservation laws. If a single particle is detected, then the consequences of its existence are prolonged to such a degree that it cannot be virtual. Virtual particles are viewed as the quanta that describe fields of the basic force interactions, which cannot be described in terms of real particles. Examples of these are static force fields, such as a simple electric or magnetic field, or any field that exists without excitations that result in its carrying information from place to place.

A virtual particle is one that does not precisely obey the $m_0^2 c^4 = E^2 - p^2 c^2$ relationship for a short time. In other words, their kinetic energy may not have the usual relationship to velocity indeed, it can be negative. The probability amplitude for them to exist tends to be canceled out by destructive interference over longer distances and times. They can be considered a manifestation of quantum tunneling. The range of forces carried by virtual particles is limited by the uncertainty principle, which regards energy and time as conjugate variables; thus virtual particles of larger mass have more limited range.

There is not a definite line differentiating virtual particles from real particles the equations of physics just describe particles (which includes both equally). The amplitude that a virtual particle exists interferes with the amplitude for its non-existence; whereas for a real particle the cases of existence and non-existence cease to be coherent with each other and do not interfere any more. In the quantum field theory view, "real particles" are viewed as being detectable excitations of underlying quantum fields. As such, virtual particles are also excitations of the underlying fields, but are detectable only as forces but not particles. They are "temporary" in the sense that they appear in calculations, but are not detected as single particles. Thus, in mathematical terms, they never appear as indices to the scattering matrix, which is to say, they never appear as the observable inputs and outputs of the physical process being modeled. In this sense, virtual particles are an artifact of perturbation theory, and do not appear in a non-perturbative treatment. There are two principal ways in which the notion of virtual particles appears in modern physics. They appear as intermediate terms in Feynman diagrams; that is, as terms in a perturbative calculation. They also appear as an infinite set of states to be summed or integrated over in the calculation of a semi-non-perturbative effect. In the latter case, it is sometimes said that virtual particles cause the effect, or that the effect occurs because of the existence of virtual particles.

There are many observable physical phenomena resulting from interactions involving virtual particles. For bosonic particles which exhibit inertial mass when they are free and "real," virtual interactions are characterized by the relatively short range of the force interaction produced by particle exchange. Examples of such short-range interactions are the strong and weak forces, and their associated field bosons. For the gravitational and electromagnetic forces, the zero rest-mass of the associated boson particle permits long-range forces to be mediated by virtual particles. However, in the case of photons, power and information transfer by virtual particles is a relatively short-range phenomenon (existing importantly only within a few wavelengths of the field-disturbance which carries information or transferred power), as for example seen in the characteristically short range of inductive and capacitive effects in the near field zone of coils and antennas (as explained in the Example 3 for the massive photons).
□

We will demonstrate that from our results, the virtual particles can be considered as real bosons in their unstable short-time states, and that such real particles can produce different physical effects, as tunneling, refraction, interference, etc.. Consequently, during the time when a particle with $m_0 = 0$ is in an unstable state, it can have collisions with another bosons, as gravitons for example, and consequently in the presence of the gravitational field, this particle may change its momentum and direction.

Based on consideration above, each particle that does not have any inertial rest-mass, it has the zero volume distribution Φ in the ordinary 3-dimensional space, with its geometric form given possibly by Dirac function. Only in this conditions the velocity of

29

one particle in the vacuum can be equal to the velocity of light which is from Einstein relativistic theory the maximal velocity for energy transportation.

This is probably only a mathematical idealization: if the 3-dimensional space is quantized as well, then the volume of these rest mass particles in their ordinary (stable) existence is just a single quant of the space different from zero.

The particles with the rest mass equal to zero, in particular situations that will be considered in the rest of this manuscript, can transform into the massive particle with a "materialized" rest-mass $\hat{m}_0$, by modifying their distribution from Dirac function, their momentum and/or direction of propagation and by diminishing their velocity of propagation. As we will see these situations can happen when, during the propagation with the velocity of light in the vacuum, it is sharply broken the spatial symmetry during this propagation when a particle reaches a surface of a dense material barrier as well. We consider as the perfect 3-dimensional space symmetry the vacuum, where each possible direction of the propagation has the same physical conditions.

The basic properties of wave-packets (that are scalar entities) described here, are common to all particles, while the different kinds of elementary particles, as photons, electrons, gravitons, etc., for example, need another properties as well. They can be done by introducing vectorial representations (as Jones vector of light polarization) or another compacted spatial dimensions, for example. From this point of view, the deterministic quantum theory presented in this section is an extension of Einstein's theory of general relativity to the microcosms of elementary particles.

2.3 External forces and particle's trajectory and geometry

From the quantum state point of view, the external "field's" forces that interfere with the particles are based on the collisions between them and a particular kind of "field's" particles with rest mass equal to zero (bosons): it is possible because the mass-particles (or excited massless particles) have a wave-packet's distribution Φ with a finite space volume, and, consequently, they may mutually collide with another particles (as bosons of this "field").

The bosons with rest-mass m_0 equal to zero can interfere between themselves only when they are excited (as, for example, when they obtain a "materialized" (8) rest-mass $\hat{m}_0 > 0$), i.e., when their space volume distribution of Φ is greater than zero, so that they can collide. During these collisions, a particle changes its total energy as well, thus its geometric form (distribution) Φ as well: in such cases in the equations (e.1) and (e.2), the component,

$$\Phi_D(\overrightarrow{\mathbf{r}}, t) = \int \int \int_{-\infty}^{+\infty} \frac{\partial B(k)}{\partial t} e^{i(k_x(x-v_x t)+k_y(y-v_y t)+k_z(z-v_z t))} dk_x dk_y dk_z,$$

becomes dominant in these unstationary (unstable) short-time intervals states, caused by the time-changing of the harmonic amplitudes $\frac{\partial B(k)}{\partial t} \neq 0$. During these collisions a particle does not change only its velocity $\overrightarrow{\mathbf{v}}$, but also its geometry (distribution) can continuously evolve. For example, as it was discussed about the changing of massive particle's geometry and resulting changing of the relativistic mass, or in "particle's explosions". This kind of explosion happens when, for instance, a massless particle

30

that propagates in the vacuum with velocity of light come up against a barrier with a small slit (where it will pass), so that this barrier interrupts drastically the space symmetry and the boundary conditions for the differential equations (e.1) and (e.2) that determine the movement and geometry of the particles.

From this point of view, the "fields" are only statistical results of the influence that high number of bosons produce to the mass-particles. For example, the electromagnetic vectorial field is a statistic result of movements of high number of photons, as it will be explained in the rest of this manuscript, and, consequently, Maxwell's lows are only the statistical lows.

The collisions of one mass-particle with bosons produce an external force $\overrightarrow{F}$, generated by the interacting bosons ("field") with this particle, that determines the propagation and trajectory of this mass-particle. In order to be able to consider the application of these external forces to the mass-particles, we need to obtain the following second-order differential equations:

Proposition 3 *The propagation and geometric form of the wave-packet* $\Psi(x, y, z, t) = \Phi(\overrightarrow{r}, t)e^{i(-\frac{\overrightarrow{p}\cdot\overrightarrow{r}}{\hbar} - \omega_0 t)}$, *of an elementary particle that propagates with velocity* $\overrightarrow{v}$, *are defined by the following second-order differential equations:*

(e.3) $(\frac{v_1}{c})^2 \triangle\Psi - \frac{1}{c^2}\frac{\partial^2\Psi}{\partial t^2} - ((\frac{\omega_p}{c})^2 + i\frac{1}{c^2}\frac{\partial\omega_p}{\partial t} - i\frac{\overrightarrow{v_1}}{c^2\hbar}\frac{\partial\overrightarrow{p}}{\partial t})\Psi - (i\frac{2\omega_p}{c^2}\overrightarrow{v_1} - \frac{1}{c^2}\frac{\partial\overrightarrow{v_1}}{\partial t})\nabla\Psi =$
$= \Upsilon_D(\overrightarrow{r}, t),$

where $\omega_p = \frac{\mathcal{E}}{\hbar} = \omega_1 + \frac{\overrightarrow{p}\cdot\overrightarrow{v_1}}{\hbar}$, *with* $\omega_1 = \frac{\partial}{\partial t}(\frac{\overrightarrow{p}\cdot\overrightarrow{r}}{\hbar} + \omega_0 t) = \frac{\overrightarrow{r}}{\hbar}\frac{\partial\overrightarrow{p}}{\partial t} + \frac{\partial}{\partial t}(\omega_0 t)$, $\overrightarrow{v_1} = \frac{\partial(\overrightarrow{v} t)}{\partial t}$, $v_1 = \sqrt{-\overrightarrow{v_1}\overrightarrow{v_1}}$, *can change in time during the propagation. The right side of (e.3)*, $\Upsilon_D(\overrightarrow{r}, t) = \frac{1}{c^2}(\overrightarrow{v_1}\nabla\Phi_D - i2\omega_1\Phi_D + \frac{\partial\Phi_D}{\partial t})e^{i(-\frac{\overrightarrow{p}\cdot\overrightarrow{r}}{\hbar} - \omega_0 t)}$, *is different from zero only in unstationary cases when* $\Phi_D(\overrightarrow{r}, t) \neq 0$.

In the stationary case, when ω_p *and* $\omega_1 = \omega_0$ *are two constants, this equation can be given in a simpler D'Alambert-like form:*

(e.4) $(\frac{v_1}{c})^2\triangle\Psi_1 - \frac{1}{c^2}\frac{\partial^2\Psi_1}{\partial t^2} = -\frac{1}{c^2}\frac{\partial\overrightarrow{v_1}}{\partial t}\nabla\Psi_1 - i\frac{\overrightarrow{v_1}}{c^2\hbar}\frac{\partial\overrightarrow{p}}{\partial t}\Psi_1,$
where $\Psi_1 = e^{i\omega_p t}\Psi(\overrightarrow{r}, t)$.

Thus, an external force $\overrightarrow{F}$ *influences a particle* Ψ *as follows:*

1. Differential equation for massive particles, non-relativistic case: then we substitute the acceleration $\frac{\partial\overrightarrow{v}}{\partial t}$ *in the equation (e.3) by the expression* $\frac{\overrightarrow{F}}{m_0}$.

2. Differential equation for relativistic case when $\hbar\omega_0$ *is a constant: then we substitute* $\frac{\partial\overrightarrow{p}}{\partial t}$ *in the equation (e.4) by* $\overrightarrow{F}$.

3. Differential equation for a massless particle (with rest mass $m_0 = 0$) *during its unstable state, when it changes the velocity* $\overrightarrow{v}$ *(with* $c > v > 0$) *and momentum* $\overrightarrow{p}$, *but does not change the total energy* $E = \hbar\omega_0$: *then we substitute* $\frac{\partial\overrightarrow{p}}{\partial t}$ *by* $\overrightarrow{F}$, ω_p *by* $\frac{\overrightarrow{r}}{\hbar}\overrightarrow{F} + \omega_0 + \frac{\overrightarrow{v_1}\cdot\overrightarrow{p}}{\hbar}$ *and* $\frac{\partial\omega_p}{\partial t}$ *by* $\frac{1}{\hbar}(\overrightarrow{r}\frac{\partial\overrightarrow{F}}{\partial t} + \overrightarrow{v_1}\overrightarrow{F} + \overrightarrow{p}\frac{\partial\overrightarrow{v_1}}{\partial t})$, *in the equation (e.3).*

Proof: From $\Psi(x, y, z, t) = \Phi(\overrightarrow{r}, t)e^{i(-\frac{\overrightarrow{p}\cdot\overrightarrow{r}}{\hbar} - \omega_0 t)}$, we obtain that:

(b.0) $\frac{\partial}{\partial t}(\nabla\Phi) = \nabla(\frac{\partial\Phi}{\partial t}) = \nabla(\overrightarrow{v_1}\nabla\Phi + \Phi_D)$ (from (a.0), proof of Proposition 1)
$= -\overrightarrow{v_1}\triangle\Phi + \nabla\Phi_D$.

(b.1) $\nabla(e^{-i\overrightarrow{p}\cdot\overrightarrow{r}/\hbar}\nabla\Phi) =$
$= (\frac{\partial}{\partial x}e_1 + \frac{\partial}{\partial y}e_2 + \frac{\partial}{\partial z}e_3)((\frac{\partial\Phi}{\partial x}e_1 + \frac{\partial\Phi}{\partial y}e_2 + \frac{\partial\Phi}{\partial z}e_3)e^{-i\overrightarrow{p}\cdot\overrightarrow{r}/\hbar}) =$
$= -(\frac{ip_x}{\hbar}\frac{\partial\Phi}{\partial x} + \frac{\partial^2\Phi}{\partial x^2} + \frac{ip_y}{\hbar}\frac{\partial\Phi}{\partial y} + \frac{\partial^2\Phi}{\partial y^2} + \frac{ip_z}{\hbar}\frac{\partial\Phi}{\partial z} + \frac{\partial^2\Phi}{\partial z^2})e^{-i\overrightarrow{p}\cdot\overrightarrow{r}/\hbar} =$

31

$$= -(-\tfrac{i\vec{\mathbf{p}}}{\hbar}\nabla\Phi + \triangle\Phi)e^{-i\vec{\mathbf{p}}\,\vec{\mathbf{r}}/\hbar}, \text{ and}$$

(b.2) $\triangle\Psi = -\nabla\nabla\Psi = -e^{-i\omega_0 t}\nabla(\nabla(\Phi e^{-i\vec{\mathbf{p}}\,\vec{\mathbf{r}}/\hbar})) =$

$$= -e^{-i\omega_0 t}\nabla((\nabla\Phi + \tfrac{i\vec{\mathbf{p}}}{\hbar}\Phi)e^{-i\vec{\mathbf{p}}\,\vec{\mathbf{r}}/\hbar}) =$$

$$= -e^{-i\omega_0 t}(\nabla(e^{-i\vec{\mathbf{p}}\,\vec{\mathbf{r}}/\hbar}\nabla\Phi) + \tfrac{i\vec{\mathbf{p}}}{\hbar}\nabla(\Phi e^{-i\vec{\mathbf{p}}\,\vec{\mathbf{r}}/\hbar})) = \quad \text{(from (b.1))}$$

$$= e^{-i\omega_0 t}((-\tfrac{i\vec{\mathbf{p}}}{\hbar}\nabla\Phi + \triangle\Phi)e^{-i\vec{\mathbf{p}}\,\vec{\mathbf{r}}/\hbar} - \tfrac{i\vec{\mathbf{p}}}{\hbar}(\nabla\Phi + \tfrac{i\vec{\mathbf{p}}}{\hbar}\Phi)e^{-i\vec{\mathbf{p}}\,\vec{\mathbf{r}}/\hbar}) =$$

$$= e^{i(-\frac{\vec{\mathbf{p}}\,\vec{\mathbf{r}}}{\hbar}-\omega_0 t)}(\triangle\Phi - \tfrac{i2\vec{\mathbf{p}}}{\hbar}\nabla\Phi + \tfrac{\vec{\mathbf{p}}\,\vec{\mathbf{p}}}{\hbar^2}\Phi) =$$

$$= e^{i(-\frac{\vec{\mathbf{p}}\,\vec{\mathbf{r}}}{\hbar}-\omega_0 t)}\triangle\Phi - \tfrac{i2\vec{\mathbf{p}}}{\hbar}(\nabla\Psi - \tfrac{i\vec{\mathbf{p}}}{\hbar}\Psi) + \tfrac{\vec{\mathbf{p}}\,\vec{\mathbf{p}}}{\hbar^2}\Psi.$$

Thus, from (b.2) we obtain that,

(b.3) $e^{i(-\frac{\vec{\mathbf{p}}\,\vec{\mathbf{r}}}{\hbar}-\omega_0 t)}\triangle\Phi = \triangle\Psi + \tfrac{i2\vec{\mathbf{p}}}{\hbar}\nabla\Psi + \tfrac{\vec{\mathbf{p}}\,\vec{\mathbf{p}}}{\hbar^2}\Psi.$ Consequently,

(b.4) $\vec{\mathbf{v_1}}\tfrac{\partial}{\partial t}\nabla\Psi = \vec{\mathbf{v_1}}e^{-i\omega_0 t}(\tfrac{\partial}{\partial t}(\nabla(\Phi e^{-i\vec{\mathbf{p}}\,\vec{\mathbf{r}}/\hbar})) - i\tfrac{\partial(\omega_0 t)}{\partial t}\nabla(\Phi e^{-i\vec{\mathbf{p}}\,\vec{\mathbf{r}}/\hbar})) =$

$$= \vec{\mathbf{v_1}}e^{-i\omega_0 t}(\tfrac{\partial}{\partial t}((\nabla\Phi + \tfrac{i\vec{\mathbf{p}}}{\hbar}\Phi)e^{-i\vec{\mathbf{p}}\,\vec{\mathbf{r}}/\hbar}) - i\tfrac{\partial(\omega_0 t)}{\partial t}(\nabla\Phi + \tfrac{i\vec{\mathbf{p}}}{\hbar}\Phi)e^{-i\vec{\mathbf{p}}\,\vec{\mathbf{r}}/\hbar}) =$$

(from (a.2) in the prof of Proposition 1)

$$= \vec{\mathbf{v_1}}e^{-i\omega_0 t}((\tfrac{\partial}{\partial t}\nabla\Phi + \tfrac{i\vec{\mathbf{p}}}{\hbar}\tfrac{\partial\Phi}{\partial t} + \tfrac{i}{\hbar}\tfrac{\partial\vec{\mathbf{p}}}{\partial t}\Phi)e^{-i\vec{\mathbf{p}}\,\vec{\mathbf{r}}/\hbar} + (\tfrac{\partial}{\partial t}e^{-i\vec{\mathbf{p}}\,\vec{\mathbf{r}}/\hbar} - i\tfrac{\partial(\omega_0 t)}{\partial t}e^{-i\vec{\mathbf{p}}\,\vec{\mathbf{r}}/\hbar})$$

$$(\nabla\Phi + \tfrac{i\vec{\mathbf{p}}}{\hbar}\Phi)) =$$

$$= \vec{\mathbf{v_1}}(\tfrac{\partial}{\partial t}\nabla\Phi + \tfrac{i\vec{\mathbf{p}}}{\hbar}\tfrac{\partial\Phi}{\partial t} + \tfrac{i}{\hbar}\tfrac{\partial\vec{\mathbf{p}}}{\partial t}\Phi - i\omega_1(\nabla\Phi + \tfrac{i\vec{\mathbf{p}}}{\hbar}\Phi))e^{i(-\frac{\vec{\mathbf{p}}\,\vec{\mathbf{r}}}{\hbar}-\omega_0 t)} =$$

(from (b.0), and (a.0) in the proof of Proposition 1)

$$= \vec{\mathbf{v_1}}(-\vec{\mathbf{v_1}}\triangle\Phi + \nabla\Phi_D + \tfrac{i\vec{\mathbf{p}}}{\hbar}(\vec{\mathbf{v_1}}\nabla\Phi + \Phi_D) + \tfrac{i}{\hbar}\tfrac{\partial\vec{\mathbf{p}}}{\partial t}\Phi - i\omega_1(\nabla\Phi + \tfrac{i\vec{\mathbf{p}}}{\hbar}\Phi))e^{i(-\frac{\vec{\mathbf{p}}\,\vec{\mathbf{r}}}{\hbar}-\omega_0 t)} =$$

(from (b.2), and (a.3) in the proof of Proposition 1)

$$= \vec{\mathbf{v_1}}(-\vec{\mathbf{v_1}}(\triangle\Psi + \tfrac{i2\vec{\mathbf{p}}}{\hbar}\nabla\Psi + \tfrac{\vec{\mathbf{p}}\,\vec{\mathbf{p}}}{\hbar}) + \tfrac{i\vec{\mathbf{p}}\,\vec{\mathbf{v_1}}}{\hbar}(\nabla\Psi - \tfrac{i\vec{\mathbf{p}}}{\hbar}\Psi) + \tfrac{i}{\hbar}\tfrac{\partial\vec{\mathbf{p}}}{\partial t}\Psi - i\omega_1(\nabla\Psi -$$

$$\tfrac{i\vec{\mathbf{p}}}{\hbar}\Psi + \tfrac{i\vec{\mathbf{p}}}{\hbar}\Psi)) + (\vec{\mathbf{v_1}}\nabla\Phi_D + i\tfrac{\vec{\mathbf{p}}\,\vec{\mathbf{v_1}}}{\hbar}\Phi_D)e^{i(-\frac{\vec{\mathbf{p}}\,\vec{\mathbf{r}}}{\hbar}-\omega_0 t)} =$$

$$= v_1^2\triangle\Psi - i(\tfrac{\vec{\mathbf{p}}\,\vec{\mathbf{v_1}}}{\hbar} + \omega_1)\vec{\mathbf{v_1}}\nabla\Psi + i\tfrac{\vec{\mathbf{v_1}}}{\hbar}\tfrac{\partial\vec{\mathbf{p}}}{\partial t}\Psi + (\vec{\mathbf{v_1}}\nabla\Phi_D + i\tfrac{\vec{\mathbf{p}}\,\vec{\mathbf{v_1}}}{\hbar}\Phi_D)e^{i(-\frac{\vec{\mathbf{p}}\,\vec{\mathbf{r}}}{\hbar}-\omega_0 t)} =$$

$$= v_1^2\triangle\Psi - i\omega_p\vec{\mathbf{v_1}}\nabla\Psi + i\tfrac{\vec{\mathbf{v_1}}}{\hbar}\tfrac{\partial\vec{\mathbf{p}}}{\partial t}\Psi + (\vec{\mathbf{v_1}}\nabla\Phi_D + i\tfrac{\vec{\mathbf{p}}\,\vec{\mathbf{v_1}}}{\hbar}\Phi_D)e^{i(-\frac{\vec{\mathbf{p}}\,\vec{\mathbf{r}}}{\hbar}-\omega_0 t)}.$$

Finally, $\tfrac{\partial^2\Psi}{\partial t^2} = \tfrac{\partial}{\partial t}(-i\omega_p\Psi + \vec{\mathbf{v_1}}\nabla\Psi + \Psi_D) =$

$$= -i\tfrac{\partial\omega_p}{\partial t}\Psi - i\omega_p\tfrac{\partial\Psi}{\partial t} + \tfrac{\partial\vec{\mathbf{v_1}}}{\partial t}\nabla\Psi + \vec{\mathbf{v_1}}\tfrac{\partial}{\partial t}\nabla\Psi + \tfrac{\partial\Psi_D}{\partial t} \quad \text{(from (e.2))}$$

$$= -i\tfrac{\partial\omega_p}{\partial t}\Psi - i\omega_p(-i\omega_p\Psi + \vec{\mathbf{v_1}}\nabla\Psi + \Psi_D) + \tfrac{\partial\vec{\mathbf{v_1}}}{\partial t}\nabla\Psi + \vec{\mathbf{v_1}}\tfrac{\partial}{\partial t}\nabla\Psi + (\tfrac{\partial\Phi_D}{\partial t} -$$

$$i\omega_1\Phi_D)e^{i(-\frac{\vec{\mathbf{p}}\,\vec{\mathbf{r}}}{\hbar}-\omega_0 t)} =$$

(from (b.4))

$$= v_1^2\triangle\Psi - \omega_p^2\Psi - i2\omega_p\vec{\mathbf{v_1}}\nabla\Psi - i\tfrac{\partial\omega_p}{\partial t}\Psi + \tfrac{\partial\vec{\mathbf{v_1}}}{\partial t}\nabla\Psi + i\tfrac{\vec{\mathbf{v_1}}}{\hbar}\tfrac{\partial\vec{\mathbf{p}}}{\partial t}\Psi + (\vec{\mathbf{v_1}}\nabla\Phi_D - i2\omega_1\Phi_D +$$

$$\tfrac{\partial\Phi_D}{\partial t})e^{i(-\frac{\vec{\mathbf{p}}\,\vec{\mathbf{r}}}{\hbar}-\omega_0 t)}.$$

Thus, by dividing with c^2 we obtain the differential equation (e.3).

When ω_p and ω_0 are two constants, then we obtain (e.4) as follows:

$$v_1^2\triangle\Psi_1 - \tfrac{\partial^2}{\partial t^2}\Psi_1 = v_1^2\triangle(\Psi e^{i\omega_p t}) - \tfrac{\partial^2}{\partial t^2}(\Psi e^{i\omega_p t}) =$$

$$= (v_1^2\triangle\Psi)e^{i\omega_p t} - \tfrac{\partial}{\partial t}((\tfrac{\partial}{\partial t}\Psi)e^{i\omega_p t} + i\omega_p\Psi e^{i\omega_p t}) =$$

$$= (v_1^2\triangle\Psi - \tfrac{\partial^2}{\partial t^2}\Psi - \omega_p^2\Psi - i2\omega_p\vec{\mathbf{v_1}}\nabla\Psi)e^{i\omega_p t} = \quad \text{(from (e.3))}$$

$$= -(\tfrac{\partial\vec{\mathbf{v_1}}}{\partial t}\nabla\Psi + i\tfrac{\vec{\mathbf{v_1}}}{\hbar}\tfrac{\partial\vec{\mathbf{p}}}{\partial t}\Psi)e^{i\omega_p t} = -\tfrac{\partial\vec{\mathbf{v_1}}}{\partial t}\nabla\Psi_1 - i\tfrac{\vec{\mathbf{v_1}}}{\hbar}\tfrac{\partial\vec{\mathbf{p}}}{\partial t}\Psi_1.$$
□

The differential equation (e.3) describes the changing (in time) of the geometry Φ of a the "corpuscular" particle (wave-packet) and its trajectory and velocity, caused by a given external force $\vec{F}$ (that is, caused by a total sum of given external "fields", as gravitational, electromagnetic, etc..).

The *Compton effect* between any two particles with their distributions $\Phi_1(x, y, z, t)$ and $\Phi_2(x, y, z, t)$, happens when there exists a space-time point (x_1, y_1, z_1, t_1) such that $\Phi_1(x_1, y_1, z_1, t_1) \cdot \Phi_2(x_1, y_1, z_1, t_1) \neq 0$. This collision of these two particles is elastic if their kinetic energies are relatively small, with the well known lows of conservation of the total energy and total momentum for elastic collisions. In the high-energy collisions we obtain a fusion of these two particles and a creation of another, as in well-known Feynman diagrams.

In the *stationary cases*, the basic equation (e.4) can be divided into following three cases:

- When the velocity $v = 0$, we obtain a simple equation:
  (e.4.0) $\frac{\partial^2 \Psi_1}{\partial t^2} = 0$, that is, $\frac{\partial^2 \Psi}{\partial t^2} = -\omega_0^2 \Psi$,
  with a simple solution, $\Psi(x, y, z, t) = \Phi(x, y, z)\mathrm{e}^{-\omega_0 t}$.

- When the velocity $\overrightarrow{\mathbf{v}}$ and the momentum $\overrightarrow{\mathbf{p}}$ are constant vectors during the propagation, different from zero, then $\overrightarrow{\mathbf{v}}_1 = \overrightarrow{\mathbf{v}}$, so we obtain D'Alambert equation where v is a constant value:
  (e.4.1) $\triangle \Psi_1 - \frac{1}{v^2}\frac{\partial^2 \Psi_1}{\partial t^2} = 0$,
  with the solution $\Psi_1(x, y, z, t) = \Phi(\overrightarrow{\mathbf{r}} - \overrightarrow{\mathbf{v}}t)\mathrm{e}^{-i\frac{\overrightarrow{\mathbf{p}}}{\hbar}(\overrightarrow{\mathbf{r}} - \overrightarrow{\mathbf{v}}t)}$, thus,
  $\Psi(x, y, z, t) = \Psi_1 \mathrm{e}^{-\omega_p t} = \Phi(\overrightarrow{\mathbf{r}} - \overrightarrow{\mathbf{v}}t)\mathrm{e}^{i(-\frac{\overrightarrow{\mathbf{p}}\cdot\overrightarrow{\mathbf{r}}}{\hbar} - \omega_0 t)}$.

- The case when the velocity $\overrightarrow{\mathbf{v}}$ with $v > 0$ and the momentum $\overrightarrow{\mathbf{v}}$ change only the direction (they are collinear vectors in each instance of time) during a propagation, but not their values, so that $-\frac{\partial(\overrightarrow{\mathbf{p}}\,\overrightarrow{\mathbf{v}})}{\partial t} = \frac{\partial(pv)}{\partial t} = 0$ and $v_1 = v$ is constant as well (this case includes the second case in Example 2 as well: when the velocity of massive particle with rest mass m_0 is $\overrightarrow{\mathbf{v}} = v(-\cos\theta e_1 + \sin\theta e_2)$, where $\theta = \nu t = \frac{v}{R}t$ is the angle w.r.t the axis x of a particle that routes around the coordinate center with constant angular velocity $\nu = \frac{v}{R}$ on circular orbit with constant radius R).
  Thus, from (e.4) we obtain an extended D'Alambert equation where $v^2 = |\overrightarrow{\mathbf{v}}\,\overrightarrow{\mathbf{v}}| > 0$ is a constant value:
  (e.4.2) $\triangle \Psi_1 - \frac{1}{v^2}\frac{\partial^2 \Psi_1}{\partial t^2} = \Theta(\overrightarrow{\mathbf{r}}, t)$,
  where $\Theta(\overrightarrow{\mathbf{r}}, t) = -\frac{1}{v^2}(\frac{\partial \overrightarrow{\mathbf{v}}}{\partial t}\nabla\Psi_1 + i\frac{\overrightarrow{\mathbf{v}}}{\hbar}\frac{\partial \overrightarrow{\mathbf{p}}}{\partial t}\Psi_1)$. Thus, we obtain a general solution $\Psi_1(\overrightarrow{\mathbf{r}} - \overrightarrow{\mathbf{v}}t) = \Phi(\overrightarrow{\mathbf{r}} - \overrightarrow{\mathbf{v}}t)\mathrm{e}^{-i\frac{\overrightarrow{\mathbf{p}}}{\hbar}(\overrightarrow{\mathbf{r}} - \overrightarrow{\mathbf{v}}t)}$.

Notice that the massive particles that does not change the total energy E during a propagation are always in the stationary states.

In all cases, the geometric form (the distribution Φ) of a particle in a given time-instance t depends on the particular boundary conditions for the differential equations as well. In the case when they are far from another massive particles (usually it can be considered if another particles are far from this particle in order of one millimeter), then Φ is symmetric w.r.t. the direction of propagation. Otherwise, the boundary conditions for these differential equations can drastically change, with the result that Φ can become enormously bigger than in normal situations, that is, they can instantaneously "explode", because the single harmonics of the Fourier representation of $\Phi(x, y, z)$ in a given time-instance t are contemporarily present in the whole 3-dimensional Euclidean space.

Let us consider the following well known example:

Example 5: One-dimensional particle in a box problem.

We assume that that the electron can move freely between two infinitely high potential barriers, along the x axes between $-\frac{a}{2}$ and $\frac{a}{2}$. The appropriate potential is $V(x) = 0$ for $-\frac{a}{2} \leq x \leq \frac{a}{2}$ and $V(x)$ equal to the infinity otherwise, that is, there are infinitely high walls at $x = -\frac{a}{2}$ and $x = \frac{a}{2}$, and the particle is trapped between them. This turns out to be quite a good approximation for electrons in a long molecule, and the three-dimensional version is a reasonable picture for electrons in metals. Consequently, the field, which is the gradient of this potential is equal to zero between these walls and directed into the box at the barriers, so that in the box we have not any field and any acceleration of the electron, so that the equation (e.4) has a simple D'Alambert form

(e.4.1) $\triangle\Psi_1 - \frac{1}{v^2}\frac{\partial^2\Psi_1}{\partial t^2} = 0$,

with the solution (we consider case when the momentum and the velociries are collinear with x axes) $\Psi_1(x, y, z, t) = \Phi(x - vt, y, z)e^{i(\frac{p}{\hbar}(x-vt))}$, where $\Phi(x - vt, y, z)$ is the geometric form of electron during its propagation (notice that this electron is at position x in the time-instance t when $x - vt = 0$), and the constant velocity v can be positive and negative (opposite direction w.r.t. the x axes).

The boundary condition at the walls is that in these points the velocity of the electron has to be equal to zero, and the stationary solution of the D'Alambert equation above has to satisfy that (after one complete oscillation, when it passes the distance $2a$), $\Psi_1(x, y, z, t) = \Psi_1(x, y, z, t + \frac{2a}{v})$, so that,

$e^{i\frac{p}{\hbar}(x-vt)} = e^{i\frac{p}{\hbar}(x-v(t+\frac{2a}{v}))}$, that is,

$\frac{p}{\hbar}(x - vt) = 2n\pi + \frac{p}{\hbar}(x - v(t + \frac{2a}{v}))$, for $n = 0, \pm1, \pm2, ..$

so that we obtain that $p = \frac{\pi\hbar}{a}n$, where $n = 0, \pm1, \pm2, ..$, exactly as in the solution of the Schrödinger equation. The positive and negative values for p corresponds to the propagation on the left or on the right (opposite direction) of this electron. Consequently we obtain a constant discrete set of possible velocities $v = \frac{p}{m_0}$ (we consider only non relativistic case when $v << c$).

The (non relativistic) kinetic energy of this electron can be one of the following discrete values:

$E = \frac{m_0v^2}{2} = \frac{p^2}{2m_0} = \frac{(\pi\hbar)^2}{2m_0a^2}n^2$, $n = 0, 1, 2, ..$

Thus, the wave-packet of this electron in the box that oscillates along the axis x, is given by:

$\Psi(x, y, z, t) = \Phi(x + la - vt, y, z)e^{i(\frac{p(x+la)}{\hbar} - \omega_0 t)}$, for $t \in [(l - \frac{1}{2})\triangle t, (l + \frac{1}{2})\triangle t]$,

or, when it propagates in opposite direction,

$\Psi(x, y, z, t) = \Phi(x - (l+1)a + vt, y, z)e^{i(\frac{-p(x-(l+1)a)}{\hbar} - \omega_0 t)}$, for $t \in [(l+1-\frac{1}{2})\triangle t, (l+1+\frac{1}{2})\triangle t]$,

where $l = 0, 1, 2, ..., \triangle t = \frac{a}{v}$, and $v \in \{n\frac{4\pi\hbar}{m_0a} \mid n = 0, 1, 2, ...\}$.

□

The D'Alambrt-like equation (e.4) is very important for the particles with the rest mass greater than zero as well, and especially to their stationary cases as well, when $\omega_p \neq 0$, as for example in the stationary orbits of electrons in the atoms. Based on the stationary solutions of the D'Alambert-like differential equation for electrons, directly from (e.4) we can derive the 3rd Bohr postulate, as follows:

Example 6: 3rd Bohr postulate for the electrons.

The electrons in the atoms have constant velocity $v > 0$ and momentum p, so that the non relativistic equation for electron's wave-packet Ψ and its propagation given by (e.4.2), can be obtained by substituting $\frac{\partial \vec{v}}{\partial t}$ with $\vec{F}/m_0$, with the radial force of the electric field $F = |\vec{F}| = \frac{Ze^2}{4\pi\epsilon_0 R^2}$, where Z is a number of protons in the atom's nucleus, e is the electrical charge of the electron, ϵ_0 the electric constant and R is the distance of the electron from the nucleus. Thus, we obtain the stationary case for the electron's propagation:

(e.5) $\quad \triangle\Psi_1 - \frac{1}{v^2}\frac{\partial^2\Psi_1}{\partial t^2} = -\frac{2}{m_0 v^2}\vec{F}\nabla\Psi_1$,

with a general solution $\Psi_1(\vec{r} - \vec{v}t) = \Phi(\vec{r} - \vec{v}t)e^{-i\frac{\vec{p}}{\hbar}(\vec{r} - \vec{v}t)}$. The stationary solutions are those for which the wave-packet Ψ_1 obtain the same identical value after every full rotation of electron around the nucleus, That is when $\Psi_1(x,y,z,t) = \Psi_1(x,y,z,t + \frac{2\pi R}{v})$, where $\frac{2\pi R}{v}$ is the time period for one complete rotation of the wave-packet Ψ_1 around the nucleus.

Thus, $\Phi(\vec{r} - \vec{v}(t + \frac{2\pi R}{v})) = \Phi(\vec{r} - \vec{v}t)$, where $\frac{2\pi R}{v}$ is the period of time for one revolution of electron around the nucleus, and we obtain that:

$\Psi_1(x,y,z,t + \frac{2\pi R}{v}) = \Phi(\vec{r} - \vec{v}(t + \frac{2\pi R}{v}))e^{-i\frac{\vec{p}}{\hbar}(\vec{r} - \vec{v}(t + \frac{2\pi R}{v}))} =$

$= \Phi(\vec{r} - \vec{v}t)e^{-i\frac{\vec{p}}{\hbar}(\vec{r} - \vec{v}t)}e^{+i\frac{\vec{p}}{\hbar}\vec{v}\frac{2\pi R}{v}} =$

$= \Psi_1(x,y,z,t)e^{-i\frac{2\pi pR}{\hbar}} = \Psi_1(x,y,z,t)$, if $\frac{2\pi pR}{\hbar} = 2n\pi$, $n = 1,2,3,..$, that is, when $pR = n\hbar$, as in the 3rd Bohr postulate. Consequently, for a given momentum of the electron, the radius R can be only one of the following values $R = n\frac{\hbar}{p}$, $n = 1,2,..$ Analogously, for a fixed radius R of an electron in a stationary states, its momentum p can have only the following discrete values $p = n\frac{\hbar}{R}$, with spatial wavelengths $\lambda = \frac{2\pi\hbar}{p} = \frac{2\pi R}{n}$, $n = 1,2,...$

Thus, for a given radius R, the wave-packet for an electron will be given by:

$\Psi(x,y,z,t) = e^{-i\omega_p t}\Psi_1 = \Phi(\vec{r} - \vec{v}t)e^{i(-\frac{\vec{p}\cdot\vec{r}}{\hbar} - \omega_0 t)} =$

$= \Phi(\vec{r} - \vec{v}t)e^{i(-\frac{n}{R}\vec{i_v}(t)\vec{r} - \omega_0 t)}$,

where, if we take the plain x,y for rotation of this electron with a center in (0,0) and initial position of electron in $t = 0$ equal to $\vec{r} = Re1 + 0e2$, then $\vec{i_v}(t) = -\cos(\frac{v}{R}t)e_1 + \sin(\frac{v}{R}t)e_2$ is the unit vector of the velocity, that is, tangent vector to the circle with radius R.

Consequently, these stationary cases correspond to the following values of ω_p:

$\omega_p = \omega_0 - \frac{pv}{\hbar} = \omega_0 - n\frac{v}{R}$, $n = 1,2,...$,

where $\omega_0 = \frac{m_0 c^2}{\hbar}$, and the stationary values for the energies $\mathcal{E} = \hbar\omega_p$ are equal to $\mathcal{E} = m_0 c^2 - n\frac{hv}{R}$, $n = 1,2,....$

As a consequence, for mass-particles we can have a number of different stationary (stable) states for ω_p.

$\square$

Chapter 3

Principal results of this theory

3.1 Massless particles: Lagrangian

Let us consider the particles with rest mass m_0 equal to zero. In that case the total energy of a particle is equal to $E = \hbar\omega_0 = p\vartheta$, where p is the momentum and $\vartheta = \frac{\omega_0}{k_0} = \frac{E}{p}$ is its *phase* velocity. Thus, in order to take in consideration the particle's velocity $\overrightarrow{\mathbf{v}}$ and to define Lagrangian $L(\overrightarrow{\mathbf{r}}, \overrightarrow{\mathbf{v}}, t)$, where for each instance of time t, $\overrightarrow{\mathbf{r}}$ is a position of particle (trajectory) and $\overrightarrow{\mathbf{v}} = \frac{\partial \overrightarrow{\mathbf{r}}}{\partial t}$ is the velocity in this time instance, we need more elaboration than in the case of the massive particles with their well defined kinetic and potential energy.

Proposition 4 *The Lagrangian of a particle with $m_0 = 0$ is equal to $L(\overrightarrow{\mathbf{r}}, \overrightarrow{\mathbf{v}}, t) = -\hbar\omega_0 - \overrightarrow{\mathbf{p}} \overrightarrow{\mathbf{v}}$, and it holds that $\frac{d\varphi_\tau}{dt} = \frac{1}{\hbar} L(\overrightarrow{\mathbf{r}}, \overrightarrow{\mathbf{v}}, t)$, where φ_τ is the phase changing of this particle along its trajectory.*
The principle of minimal action means that this phase has to obey the extremum condition $\delta\varphi_\tau = \frac{1}{\hbar}\delta(\int_{t_1}^{t_2} L(\overrightarrow{\mathbf{r}}, \overrightarrow{\mathbf{v}}, t)dt)$, for $t_2 > t_1$.
The Fermat's principle in optics corresponds to this minimal action for particles that does not change their total energy $E = \hbar\omega_0$ during the propagation in a given interval of time (t_1, t_2).

Proof: The Lagrangian for such a particle with $m_0 = 0$, valid in all cases, has to be valid also when $E - pv > 0$ (it can happen only when their velocity v is less than the velocity c of light in the vacuum) and when the space explosion of a particle (from Dirac function of its stable state) happens, with a consequence (8) that they have an amount of "materialized" rest-mass $\widehat{m_0}$.
Thus, in such particular conditions, when $E - pv$ is enough big value, that is, when $E - pv > \mathcal{E}_0$ where $\mathcal{E}_0 = E(1 - \frac{v}{c})$, we obtain a "materialized" rest-mass $\widehat{m_0}$ for this unstable particle during its space-explosion, defined by (8), $\widehat{m_0} = \frac{1}{c^2}\sqrt{E^2 - (pc)^2} > 0$, such that the correspondent relativistic mass is equal to $m = \frac{E}{c^2} = \frac{m_0}{\sqrt{1-(v/c)^2}}$, and the momentum $p = \frac{1}{c^2}\sqrt{E^2 - (\widehat{m_0}c^2)^2} = mv$.
From the fact that the vector momentum $\overrightarrow{\mathbf{p}}$ and the vector velocity $\overrightarrow{\mathbf{v}}$ of particle are collinear, then $E + \overrightarrow{\mathbf{p}} \overrightarrow{\mathbf{v}} = E - pv = E - mv^2 = (E - \frac{mv^2}{2}) - \frac{mv^2}{2} \geq 0$ is, at each instance of time, the difference between the total energy E and "momentum" energy pv in that instance of time.
Thus, if we consider that $E - \frac{mv^2}{2}$ is a *potential* energy (difference between the total energy and kinetic energy) then $-(E - pv)$ is the difference between the kinetic and potential energy, that is, it is equal to the Lagrangian $L(\overrightarrow{\mathbf{r}}, \overrightarrow{\mathbf{v}}, t) = -(E - pv) = -\hbar\omega_0 - \overrightarrow{\mathbf{p}} \overrightarrow{\mathbf{v}}$.
Thus, from the fact that the particle's velocity is equal to $\overrightarrow{\mathbf{v}} = \frac{d\overrightarrow{\mathbf{r}}_\tau}{dt}$, where $d\mathbf{r}_\tau$ is an infinitesimal vector tangent on the particle's trajectory, we obtain:
$L(\overrightarrow{\mathbf{r}}, \overrightarrow{\mathbf{v}}, t) = -\hbar\omega_0 - \overrightarrow{\mathbf{p}} \overrightarrow{\mathbf{v}} = -\hbar\omega_0 - \overrightarrow{\mathbf{p}} \frac{d\overrightarrow{\mathbf{r}}_\tau}{dt} = \frac{\hbar}{dt}(-\frac{\overrightarrow{\mathbf{p}}}{\hbar}d\overrightarrow{\mathbf{r}}_\tau - \omega_0 dt) = \hbar\frac{d\varphi_\tau}{dt}$,
where φ_τ is the phase changing of this particle along its trajectory (notice that $\Psi(\overrightarrow{\mathbf{r}}, t) = \Phi(\overrightarrow{\mathbf{r}}, t)e^{i(-\frac{\overrightarrow{\mathbf{p}} \overrightarrow{\mathbf{r}}}{\hbar} - \omega_0 t)}$ where $-\frac{\overrightarrow{\mathbf{p}} \cdot \overrightarrow{\mathbf{r}}}{\hbar} - \omega_0 t$ is its phase φ).
Let us denote by a and b the points of a photon in time instant t_1 and t_2 respectively. Then, in the case when $E = \hbar\omega_0$ is constant in this interval of time, we have that:
$\delta\varphi_\tau = \frac{1}{\hbar}\delta(\int_{t_1}^{t_2} L(\overrightarrow{\mathbf{r}}, \overrightarrow{\mathbf{v}}, t)dt) = \frac{1}{\hbar}\delta(\int_{t_1}^{t_2}(-E - \overrightarrow{\mathbf{p}} \overrightarrow{\mathbf{v}})dt) = \frac{1}{\hbar}\delta(\int_{t_1}^{t_2} -\overrightarrow{\mathbf{p}} \overrightarrow{\mathbf{v}}dt) =$

$\frac{1}{\hbar}\delta(\int_a^b - \overrightarrow{\mathbf{p}}\,d\overrightarrow{\mathbf{r}_\tau}) = \delta(\int_a^b k_0 dl) = \frac{\omega_0}{c}\delta(\int_a^b n dl)$,

where $n = \frac{c}{\vartheta} = \frac{ck_0}{\omega_0}$ is the refractive index and dl infinitesimal length of photon's trajectory (i.e., infinitesimal optical path length), so that $\delta(\int_a^b n dl)$ is obtained Fermat's principle.

It means that the refraction of photons and any massless particle is obtained when the total energy of a particle does not change during a propagation.

$\square$

Thus, the refraction of a single photon is equivalent to the refraction of the light (wave-optics) given by Fermat's principle. The minimal action of the Lagrangian of a single photon is a quantistic equivalence of the Fermat's principle in wave-optics. In what follows, we will try to analyze the dynamics of a single massless particle during this process of refraction, and to show its quantum mechanisms in Example 7 by deriving the well-known Snell's low for the refraction of light.

Notice that derivation of least action does not use any kind of 'phase wave' as in [4], where De Broglie did not give any explanation about the mining of this 'phase wave': but he was guided by "the idea that a corpuscle and its phase wave are not separate physical realities". In fact, as resulting from our deterministic theory, a particle is defined as a Minkowski's time-space wave-packet $\Psi(\overrightarrow{\mathbf{r}}, t) = \Phi(\overrightarrow{\mathbf{r}}, t)e^{i(-\frac{\overrightarrow{\mathbf{p}}\cdot\overrightarrow{\mathbf{r}}}{\hbar} - \omega_0 t)}$, where following de Broglie, $\Phi(\overrightarrow{\mathbf{r}}, t)$ is a 'corpuscle'(real function) and $e^{i(-\frac{\overrightarrow{\mathbf{p}}\cdot\overrightarrow{\mathbf{r}}}{\hbar} - \omega_0 t)}$ is its complex phase wave. Here we used only real particle's properties, as total energy, its real velocity and momentum. Moreover, in the next Example 7, we will use the real particle's velocity in order to explain the refraction (a changing of the direction of a propagation) of photon during its propagation in nonhomogeneous medium, instead of the Fermat's principle of geometrical (wave) optics derived from Proposition 4 above. Unstable particles tends naturally to return in their stable state (stationary) when the external solicitations that cause their unstable states are eliminated.

Thus in this stable state (ω_p and ω_0 are two constants) when a massless particle with $v > 0$ have no acceleration ($\frac{\partial \overrightarrow{v}}{\partial t} = 0$), the differential equation (e.4) becomes equal to the well known Jean le Rond d'Alembert equation with $\Psi_1 \equiv \Psi = \Phi(\overrightarrow{\mathbf{r}} - \overrightarrow{\mathbf{v}}t)e^{-i\frac{\overrightarrow{\mathbf{p}}}{\hbar}(\overrightarrow{\mathbf{r}} - \overrightarrow{\mathbf{v}}t)}$:

(e.4.1) $\triangle\Psi - \frac{1}{v^2}\frac{\partial^2\Psi}{\partial t^2} = 0$,

where from (1),

$\Phi(\overrightarrow{\mathbf{r}} - \overrightarrow{\mathbf{v}}t) = \Phi(x - v_x t, y - v_y t, z - v_z t) = $

$\int\int\int_{-\infty}^{+\infty} B(k)e^{i(k_x(x-v_x t)+k_y(y-v_y t)+k_z(z-v_z t))}dk_x dk_y dk_z$,

with the amplitudes of the harmonics $B(k)$ that depend on the concrete boundary conditions.

This stable state when this particle propagates in the free space (without local discontinuities in any direction w.r.t. the current position of this particle), thus in complete space symmetry, corresponds to the boundary conditions for the differential equation (e.4.1) of this wave-packet that results in their solution equal to the Dirac function $\Phi(\overrightarrow{\mathbf{r}} - \overrightarrow{\mathbf{v}}t) = \delta(\overrightarrow{\mathbf{r}} - \overrightarrow{\mathbf{v}}t)$, when $B(k)$ are constant and equal to $\frac{1}{(2\pi)^3}$ (it determines the fact that its geometric distribution Φ is the Dirac function).

But in the absence of the complete local space symmetry around this particle, for example when this particles passes very close to some spatial discontinuity (as, for example, when it passes from the vacuum into another isotropic media as air, water, glass, etc..), the boundary conditions in the very-close presence of the surface that divides two different mediums of propagation, are very different from those in the free symmetric space. This local spatial symmetry is sharply broken when this particle riches this surface (its atoms of the medium): the different boundary conditions for differential equation (e.3) change *instantaneously* the amounts of amplitudes of harmonics $B(k)e^{i(k_x(x-v_xt)+k_y(y-v_yt)+k_z(z-v_zt))}$, because in these moments $\frac{\partial B(k)}{\partial t}$ become generally different from zero and the component $\Upsilon_D(\vec{r},t)$ in (e.3) becomes significant, which can produce a spatial explosion of the particle (a distribution Φ changes from Dirac zero-volume state in the state where its volume can be very big). This excited unstable state of this particle, when it is involved in this space explosion with a "materialized" rest-mass $\hat{m}_0 > 0$) is usually localized in very small interval of time $\triangle t \approx \frac{\triangle l}{v} \approx 0$, where $\triangle l$ is very small width (w.r.t. the direction of its propagation) of this particle and v very high velocity of its propagation. When this particle passes completely this surface (that has broken locally the space symmetry) it becomes again stable state particle with Dirac function (zero-volume) geometric form. Thus, after this interval of time $\triangle t$, this particle propagates again in the new local space-symmetry (of this medium) w.r.t. its trajectory, and, consequently, returns in its stable state.

During this interval of time $\triangle t$ the particle can change its velocity and momentum (for example, in the simplest case their direction), without changing its total energy E as well (as in the case of the refraction when the Fermat's principle is valid, as obtained in Proposition 4).

The instantaneous change of coefficients $B(k)$, when is broken the local space symmetry around the particle, makes its space distribution Φ present *instantaneously* in the local space on both sides of the surface that divides two mediums, but without any transportation of its energy (thus, by conserving the special relativity principle where the maximal velocity for the transportation of particle's energy is finite and equal to the velocity of light). It explains the quantum effects, when a single particle passes through a small slit of a metal surface with another distant slit (the *double-slit* Young's experiment): the particle will "see" this distant slit as well in this unstable, non-zero volume state, in the way that Φ will be present in both slits *contemporarily and instantaneously*, so that it will experiment interference effects typical for plain waves. Notice that the same effect of geometric "explosion" happens for the massive particles as well, when the local boundary conditions becomes different from the propagation of these particles in the free space of the vacuum. It is clear, for example, that a number of photons that are moving from the vacuum into an isotopic medium can be reflected as well (Compton effect with some massive particle in the surface of this material medium), or absorbed inside this medium in an high-energy collision with some massive particle inside this medium. Thus, statistically, the propagation of a stream of photons will have well known effects in physical optics.

Let us consider the case when the total energy of this particle does not change. We consider that in the infinitesimal time interval $\triangle t$ of its spatial explosion, without enormously strong local external fields this excited particle will not change its total energy, that can be caused only by collisions with another locally presented bosons. The most known effects as refraction, wave propagation interferences, etc..in the case of the single photons does not change their total energy, thus the boundary conditions for the differential equation (e.3)) for the wave-packet geometric form Φ of the excited photon (unstable state) has to satisfy the condition that $\Phi(x, y, z, t) = 0$ in all local points (x, y, z, t), around its position immediately before its space-geometric explosion from the Dirac (stable) function, where are localized another massive particles (with rest mass greater than zero) with their space forms $\Phi_i, i = 1, 2, 3, ..$ These boundary conditions avoid the collisions (Compton effects) of this excited non-zero-volume particle with these local massive particles, because such collisions would necessarily modify not only the momentum but its total energy as well.

As in the case of Lagrangian $L(\overrightarrow{\mathbf{r}}, \overrightarrow{\mathbf{v}}, t)$, where $(\overrightarrow{\mathbf{r}}, t)$ are the space-time coordinates *of the path* of a particle, we can obtain the reduction of the differential equation (e.3) to the differential equation valid only on points $(\overrightarrow{\mathbf{r}}, t)$ of the path of this particle: such a difference equation will determine only the points of the path and the phase of this particle in each point of the path. Typical application for such a reduction is when we are considering the propagation path of a massless particle such that its total energy $E = \hbar\omega_0$ does not change and the *product* of momentum and velocity $\overrightarrow{\mathbf{p}}\,\overrightarrow{\mathbf{v}}$ does not change as well, for instance the optical path of photons in a given refractive medium. In this case we will require for $\omega_p(\overrightarrow{\mathbf{r}}, t)$ to be constant only on the path-points $(\overrightarrow{\mathbf{r}}, t)$, and not for each pair $(\overrightarrow{\mathbf{r}}, t)$ (as in the requirement for stationary states), that is:

(e.6) $\quad \omega_p(\overrightarrow{\mathbf{r}}, t) = \frac{\overrightarrow{\mathbf{r}}}{\hbar}\frac{\partial\overrightarrow{\mathbf{p}}}{\partial t} + \frac{\partial\omega_0 t}{\partial t} + \frac{\overrightarrow{\mathbf{p}}}{\hbar}\frac{\partial\overrightarrow{\mathbf{v}}}{\partial t} = \frac{1}{\hbar}(\hbar\omega_0 + \overrightarrow{\mathbf{p}}\,\overrightarrow{\mathbf{v}} + \overrightarrow{\mathbf{r}}\frac{\partial\overrightarrow{\mathbf{p}}}{\partial t} + t\,\overrightarrow{\mathbf{p}}\frac{\partial\overrightarrow{\mathbf{v}}}{\partial t}) = $

(from the fact that $0 = \frac{\partial\overrightarrow{\mathbf{p}}\,\overrightarrow{\mathbf{v}}}{\partial t} = \overrightarrow{\mathbf{p}}\frac{\partial\overrightarrow{\mathbf{v}}}{\partial t} + \overrightarrow{\mathbf{v}}\frac{\partial\overrightarrow{\mathbf{p}}}{\partial t}$)

$= \frac{1}{\hbar}(\hbar\omega_0 + \overrightarrow{\mathbf{p}}\,\overrightarrow{\mathbf{v}} + \frac{\partial\overrightarrow{\mathbf{p}}}{\partial t}(\overrightarrow{\mathbf{r}} - \overrightarrow{\mathbf{v}}t)) = \frac{1}{\hbar}(\hbar\omega_0 + \overrightarrow{\mathbf{p}}\,\overrightarrow{\mathbf{v}})$,

because $\overrightarrow{\mathbf{r}} - \overrightarrow{\mathbf{v}}t = 0$ when $(\overrightarrow{\mathbf{r}}, t)$ is a point of path. Let us consider, for example the refraction of a photon when it passes from one into another medium: before the impact on the dividing boundary it is in a stable situation with constant momentum $\overrightarrow{\mathbf{p}}$ and velocity $\overrightarrow{\mathbf{v}}$. Let $(\overrightarrow{\mathbf{r}} = 0, t = 0)$ be the point of trajectory on this dividing boundary, so that after the refraction this particle becomes stable again, with a new constant values of the momentum $\overrightarrow{\mathbf{p}_1}$ and velocity $\overrightarrow{\mathbf{v}_1}$, such that $pv = -\overrightarrow{\mathbf{p}}\,\overrightarrow{\mathbf{v}} = -\overrightarrow{\mathbf{p}_1}\overrightarrow{\mathbf{v}_1} = p_1 v_1$. It is easy to see that both $\overrightarrow{\mathbf{r}} - \overrightarrow{\mathbf{v}}t = 0$ for $t < 0$, and $\overrightarrow{\mathbf{r}} - \overrightarrow{\mathbf{v}_1}t = 0$ for $t > 0$ hold for the points $(\overrightarrow{\mathbf{r}}, t)$ of the particle's path.

Thus, in this case, for all points of the path we obtain that $\omega_p = \omega_0 + \frac{\overrightarrow{\mathbf{p}}\,\overrightarrow{\mathbf{v}}}{\hbar}$ is a constant value. Consequently the general equation (e.3) can be reduced to D'Alambert equation (e.4) applicable only to path-points.

Consequently, the solution of (e.4) in this case is given by $\Psi(\overrightarrow{\mathbf{r}} - \overrightarrow{\mathbf{v}}t) = \Psi(\overrightarrow{\mathbf{r}} - \overrightarrow{\mathbf{v}}t)\mathrm{e}^{-i\omega_p t} = \Phi(0)\mathrm{e}^{i\varphi}$, where $\Phi(0)$ and $\varphi = -\frac{\overrightarrow{\mathbf{p}}\,\overrightarrow{\mathbf{r}}}{\hbar} - \omega_0 t$ are the value of particle's distribution and the phase of this particle in a path-point $(\overrightarrow{\mathbf{r}}, t)$, relatively.

Remark: Notice that in order to define the path and the phase of a particle in this example, we do not need that this particle remains stable in all points of its path, that is, we have that generally $\omega_p((\overrightarrow{\mathbf{r}}, t)) \neq 0$ and it varies from the point to point outside of the particle path. So that for all other points $(\overrightarrow{\mathbf{r}}, t)$ different from the particle's trajectory (but close to it for instance) the distribution of the particle Φ can vary as for general unstable cases (as a particle "explosion") and is defined by more complex differential equation (e.3).

$\square$

Thus, it is possible to have a deceleration of a massless particle with the contemporary presence of augmentation of its momentum, in particular when $\hbar\omega_p = E - pv$ is constant (does not change in time). Consequently, if the totaly energy E remains unchanged, than pv is constant during the interval of time $\triangle t$ as well. In that case we will have that the product vp of the velocity and momentum of this particle before its excitation, and $v_1 p_1$ (after its excitation) are equal, with $v > v_1, p < p_1$ or viceversa, and/or with a change of the direction of the propagation.

This particular case will be described in the following example:

Example 7: Snell's low for the refraction of light.

In optics and physics, Snell's law (also known as Descartes' law, the SnellDescartes law, and the law of refraction) is a formula used to describe the relationship between the angles of incidence and refraction, when referring to light or other waves passing through a boundary between two different isotropic media, such as water and glass. The law says that the ratio of the sines of the angles of incidence and of refraction is a constant that depends on the media. The law follows from the boundary condition that a wave be continuous across a boundary, which requires that the phase of the wave be constant on any given plane, resulting in

(11) $\quad \frac{\sin\theta_1}{\sin\theta_2} = \frac{\vartheta_1}{\vartheta_2} = \frac{\lambda_1}{\lambda_2} = \frac{n_2}{n_1}$,

where $\theta_1 < \frac{\pi}{2}$ and θ_2 are the angles from the normal of the incident and refracted waves, respectively; the ϑ_1, λ_1 and ϑ_2, λ_2 are the phase velocity and wave-length of light of the incident and refracted waves respectively, and $n = \frac{c}{\vartheta}$ is the refractive index of a medium.

But this low does not explain *why* in the case when incidence is from a first isotropic medium and refraction in some isotropic media with greater refractive index, the velocity in this second isotropic media is lower than that in the vacuum (where velocity is equal to c), and what is happening with a *single photon* to change its direction of propagation and velocity by preserving its total energy.

In what follows we will consider a single photon that propagates in the plain xy where x is the axis that coincides with the surface that divides these two isotopic media, and y axis that corresponds to the normal to this dividing surface (where $z = 0$) with direction from the first (incident) to the second media. Thus, the velocity of a photon $\overrightarrow{\mathbf{v}} = v_x e_1 + v_y e_2$, with its orthogonal components $v_x = v\sin\theta_1 > 0, v_y = v\cos\theta_1 > 0$, and momentum $\overrightarrow{\mathbf{p}} = p_x e_1 + p_y e_2$, with its orthogonal components $p_x = p\sin\theta_1 > 0, p_y = p\cos\theta_1 > 0$, in its stable state satisfy the condition that particle's total energy E, and $\mathcal{E} = E + \overrightarrow{\mathbf{p}}\,\overrightarrow{\mathbf{v}} = E - pv$ (and consequently, ω_p) are constant.

In the time instance $t_0 = 0$ when this stable photon passes the boundary between these two isotropic media, it becomes non-zero volume particle, because of the fact that, in the surface that divide these two media, the local space-symmetry is broken, with the resulting smaller velocity $v' \leq v$ of this non-zero-volume photon (that changes its Dirac function of distribution into an non-zero volume distribution of Φ) but without changing its total energy E.

The phenomena of changing the direction of its propagation can be explained intuitively as follows. Let Φ of this non-zero-volume photon be a disc, symmetric w.r.t. the direction of propagation. Then, when it begins to pass the surface that divide these two mediums, its inferior part that penetrates into this medium finds greater propagation

42

resistance than its upper part that still propagates in the first medium with velocity v. Thus, this difference in the resistance of the propagation causes a momentum for its inertial mass $m = \frac{E}{c^2}$, so that this particle change its direction of propagation caused by this resistance of the medium. This medium's resistance to the particle's propagation can be formally expressed as a constant force orthogonal to the surface of this medium in the point of penetration of a photon into this medium, that is, to the constant deceleration $A = -\frac{\partial v'}{\partial t}$ of this particle during the interval $\triangle t$.

During all interval $\triangle t$ of this unstable state, we have that ω_p and E are constant,i.e., $\frac{\partial \omega_p}{\partial t} = 0$ on the particle's *path* (as explained previously in (e.5)).

Let us suppose that during this interval of time this deceleration is constant, thus $A = -\frac{v'(\triangle t) - v'(t_0)}{\triangle t} = \frac{v' - v'(\triangle t)}{\triangle t}$ (as far as I know it is one of the original Newton's ideas for "corpuscular" light). Thus, from the fact that pv' is constant we obtain that, $-p\frac{\partial v'}{\partial t} = v'\frac{\partial p}{\partial t}$.

Consequently, $dp = \frac{p}{v'}A dt$, that is, by integration, for the orthogonal component (to the dividing surface between these two mediums) of the momentum we have that (here $pv' = E - \mathcal{E}$ is constant during the interval $\triangle t$)

$p_{1y} = p_y(\triangle t) = p_y + A \int_0^{\triangle t} \frac{p}{v'} dt = p\cos\theta_1 + \frac{A}{pv'} \int_0^{\triangle t} p^2 dt = p\cos\theta_1 + \frac{A}{pv'} \int_0^{\triangle t} (p_x^2 + p_y^2) dt = p\cos\theta_1 + \frac{A}{pv'} \int_0^{\triangle t} ((p\sin\theta_1)^2 + (p\cos\theta_1 + at)^2) dt$, where a is a constant in Taylor's polynomial approximation at time instance $t_0 = 0$ for all $t_0 \leq t \leq \triangle t$, and $\triangle t \approx 0$. Thus we obtain

(c.1) $p_{1y} \approx p\cos\theta_1 + \frac{A}{pv'} p^2 \triangle t = (1 + K)p\cos\theta_1 = (1 + K)p_y > p_y$,

where $K = \frac{A\triangle t}{v'\cos\theta_1} = \frac{v' - v'(\triangle t)}{v'\cos\theta_1} > 0$, while the components of the momentum orthogonal to F (tangent to the surface of this medium) are unchanged, that is

(c.2) $p_{1x} = p_x = p\sin\theta_1$.

Consequently, the angle θ_2 of refraction is obtained from $\tan\theta_2 = \frac{p_{1x}}{p_{1y}}$, thus

(c.3) $\theta_2 = \arctan(\frac{\tan\theta_1}{1+K}) < \theta_1$.

At the end of the time interval $\triangle t$, when whole excited photon has passed the surface of the medium, it continues to propagate again into symmetric isotropic space, thus, it becomes again stable with Dirac function geometric distribution and with the same constant value $\omega_p = (E - p_1v_1)/\hbar$. Thus $p_1v_1 = pv$, so that its final velocity in the medium is equal to:

(c.4) $v_1 = \frac{vp}{p_1} = v\frac{p_x/\sin\theta_1}{p_{1x}\sin\theta_2} = v\frac{\sin\theta_2}{\sin\theta_1} < v$,

that is, we obtained the well known Snell's low

(11) $\frac{\sin\theta_1}{\sin\theta_2} = \frac{p_1}{p} = \frac{v}{v_1} = \frac{\lambda}{\lambda_1}$,

(from the fact that the total energy does not change, we have that $\frac{\sin\theta_1}{\sin\theta_2} = \frac{p_1}{p} = \frac{\vartheta}{\vartheta_1}$ as well), and the wave-packet of the stable photon after $\triangle t$ in this medium is

$\Psi(x, y, z, t) = \delta(\overrightarrow{\mathbf{r}} - \overrightarrow{\mathbf{v}}_1 t)e^{-i\frac{\overrightarrow{E}_1}{\hbar}(\overrightarrow{\mathbf{r}} - \overrightarrow{\mathbf{v}}_1 t)}$,

with the same unchanged total energy $E = p\vartheta = p_1\vartheta_1$.

$\square$

Remark: we assume that pv is constant during this transition and the total energy $E = p\vartheta$ as well. Is it real, and what are the consequences? If the total energy E does not change, then the particle's mass $m = \frac{E}{c^2}$ does not change, so that for massless particles we have no the dependence of momentum on the velocity by (*) $p = mv = \frac{E}{c^2}v$

(that is, the relationship $\vartheta v = c^2$). In fact, if $p = mv$ would hold for the massless particles as well, then we would have that $pv = \frac{E}{c^2}v^2$, so if it is constant by assumption than the velocity of particle would be constant. This is in contrast with experiments, where the velocity of a particle (i.e., the group velocity) slow down in the medium as glass for example. Consequently, we have that for the massless particles $p \neq mv = \frac{E}{c^2}v$, so that for the constant energy, the momentum p *is not* proportional to the velocity v, and we can have the refraction effects when the momentum grows up while the velocity contemporary slows down by conserving their product pv constant in time.

Notice that another proof for $p \neq mv$ comes from the refraction experiments expressed by (11): if the phase velocity $\vartheta = c$, then for $\theta_1 > \theta_2$ we have that the phase velocity in the medium $\vartheta_1 < c$, and consequently, from the condition (*) we would obtain the particle's velocity $v_1 > c$ what is impossible, so that $v_1 \vartheta_1 \neq c^2$ in a medium with refractive index different from 1.

$\square$

We can verify that the *group velocity* $\frac{\partial \omega_0}{\partial k_0} = \frac{\partial E}{\partial p}$ of a massless particle corresponds to the particle's velocity (here we can not use the De Broglie demonstration given in [4] for the massive particles, because for the massless particles, as explained in the remark above, it generally does not hold that $p = mv$ and $m = \frac{m_0}{\sqrt{1-\beta^2}}$), for any particle that propagates with a given constant phase velocity $\vartheta = \frac{E}{p} = \frac{\omega_0}{k_0}$, so that its total energy $E = \hbar \omega_0 = \hbar \omega(k_0)$ is proportional to its momentum $p = \hbar k_0$, that is, when

(**) $\frac{\partial(\hbar \omega_0)}{\partial p} = \frac{\partial(\hbar \omega(k))}{\partial p}/_{k=k_0} = 0$.

In fact from (0) we have that $\hbar(\omega(k) - \omega_0) = \hbar \omega(k) - E = -\hbar \vec{v}(\vec{k} - \vec{k_0}) = -\hbar \vec{v}\vec{k} - vp$ is valid for *all* free variable vectors $\vec{k}$ and, consequently, $\frac{\partial(\hbar \omega(k))}{\partial p} - \frac{\partial E}{\partial p} = -\hbar \vec{k} \frac{\partial \vec{v}}{\partial p} - \frac{\partial(vp)}{\partial p} = -\frac{\partial(v)}{\partial p}(\hbar \vec{k}\, \vec{i_v} + p) - v$, i.e.,

$\frac{\partial E}{\partial p} - v = \frac{\partial(\hbar \omega(k))}{\partial p} + \frac{\partial(v)}{\partial p}(\hbar \vec{k}\, \vec{i_v} + p)$.

Notice that the left part of this equation is independent on the free variable $k = \vec{k}$, while the right part is dependent on it, so we have to show that they are constant and equal to zero, in order to be valid equation for all free variables k.

It is enough to demonstrate that the right part is equal to zero for a particular value k, and let us take that $k = k_0$. Then,

$[\frac{\partial(\hbar \omega(k))}{\partial p} + \frac{\partial(v)}{\partial p}(\hbar \vec{k}\, \vec{i_v} + p)]/_{k=k_0} = \frac{\partial(\hbar \omega(k))}{\partial p}/_{k=k_0} + \frac{\partial(v)}{\partial p}(\hbar \vec{k_0}\, \vec{i_v} + p) = \frac{\partial(\hbar \omega(k))}{\partial p}/_{k=k_0} + \frac{\partial(v)}{\partial p}(-\hbar k_0 + p) = \frac{\partial(\hbar \omega(k))}{\partial p}/_{k=k_0} = 0$, from (**).

Consequently, we obtain that $\frac{\partial E}{\partial p} = v$. Thus, the group velocity corresponds to the velocity v of a massless particle, and the dependence of the particle's velocity v and its phase velocity ϑ is given by the well-known Rayleigh low,

$v = \frac{\partial E}{\partial p} = \frac{\partial(p\vartheta)}{\partial p} = \vartheta - \lambda \frac{\partial \vartheta}{\partial \lambda} = \vartheta(1 + \frac{\lambda}{n}\frac{\partial n}{\partial \lambda}) \leq c$, or $\vartheta = v(1 + \frac{\lambda}{n}\frac{\partial n}{\partial \lambda})^{-1}$,

where $n = \frac{c}{\vartheta}$ is the refractive index. It can be seen that the velocity v of particle is greater than its phase velocity only for anomaly dispersion when $\frac{\partial n}{\partial \lambda} > 0$, but in that case we have the absorption of particles in the passing mediums, so that the fundamental special-relativity condition $v \leq c$ is valid for these mediums as well.

44

3.2 Stream of particles: statistical meaning of Schrödinger equation

It is well known that the relativistic version of the Schrödinger equation for the elementary particles is postulated by the following Klein-Gordon second-order differential equation:

(e.7) $\quad \triangle \psi - \frac{1}{c^2}\frac{\partial^2}{\partial t^2}\psi = (\frac{\omega_0}{c})^2\psi,$

whose solution can be given by $\psi = \varphi(x, y, z, t)e^{-i\omega_0 t}$.

The solution for a propagation of the elementary particles in the stationary cases, when ω_0 and ω_p are constant in time, presented previously, can be postulated in the similar way, by using the particle's second-order D'Alambert-like differential equation:

(e.4) $\quad (\frac{v_1}{c})^2\triangle\Psi_1 - \frac{1}{c^2}\frac{\partial^2\Psi_1}{\partial t^2} = -\frac{1}{c^2}\frac{\partial\overrightarrow{v_1}}{\partial t}\nabla\Psi_1 - i\frac{\overrightarrow{v_1}}{c^2\hbar}\frac{\partial\overrightarrow{p}}{\partial t}\Psi_1,$

where $\Psi_1 = e^{i\omega_p t}\Psi(\overrightarrow{r}, t)$,

with $\Psi_1 = e^{i\omega_p t}\Psi$, where $\Psi(x, y, z, t)$ describes an elementary particle that propagates with a velocity $\overrightarrow{v}$ and a momentum $\overrightarrow{p}$, and, possibly, with an acceleration $\frac{\partial\overrightarrow{v}}{\partial t}$, such that $\omega_p = \frac{E + \overrightarrow{p}\,\overrightarrow{v}}{\hbar} = \frac{E - pv}{\hbar}$, where E is the potential energy of this particle (equal to $m_0 c^2$) if $m_0 > 0$, and total energy otherwise.

Then, the solution of (e.4) is $\Psi_1(\overrightarrow{r} - \overrightarrow{v}t) = \Phi(\overrightarrow{r} - \overrightarrow{v}t)e^{i\frac{\overrightarrow{p}}{\hbar}(\overrightarrow{r} - \overrightarrow{v}t)}$, where real function $\Phi(\overrightarrow{r} - \overrightarrow{v}t)$, at any fixed time instance t is a geometric distribution of this particle in that moment. It depends on the boundary conditions for the differential equation (e.4) for this particle in this given time-instance t. If this particle propagates in the vacuum, than Φ is symmetric w.r.t the direction of propagation, but if in a given moment t this particle propagates nearly to another massive particles, than the boundary conditions change and depend on the spatial presence (their distributions) of these particles. Consequently, the geometric distribution of a massive particle can change (see the equation (e.3) for the general case) also when it propagates with constant velocity $\overrightarrow{v}$, in presence of another massive particles.

This fact is very important when a particle that propagates in the vacuum, encounter some material obstacle, so that Φ, in the momentum when is broken this space symmetry, can drastically change and increment its volume, in a short interval of time. It is important to denote that such a changing of the geometric distribution is instantaneous in all points of the space, and has no any constraint as, for example, the maximal velocity of energy transportation, that is equal to the velocity of light from relativistic theory of Einstein.

From the fact that $\Psi_1 = e^{i\omega_p t}\Psi$, we obtain that any elementary particle has the following wave-packet geometric form (that was previously postulated in (1)):

$\Psi(x, y, z, t) = \Phi(\overrightarrow{r} - \overrightarrow{v}t)e^{i(-\frac{\overrightarrow{p}\,\overrightarrow{r}}{\hbar} - \omega_0 t)}$

In the case when the velocity of propagation is constant, equal to zero, then we obtain the solution $\Psi(x, y, z, t) = \Phi(x, y, z)e^{-i\omega_0 t}$, where $\omega_0 = \frac{E}{\hbar}$ and E is the potential energy of this particle.

There is the following relationship between Klein-Gordon and our differential equations:

Proposition 5 *The Klein-Gordon equation corresponds to the differential equation (e.4) of an elementary particle if its velocity of propagation is equal to zero.*

Proof: In the case when $\vec{v} = 0$ and $\frac{\partial \vec{v}}{\partial t} = 0$, so that $\omega_p = \omega_0$ is constant, then, from (e.4), we obtain that:

$$0 = \frac{\partial^2 \Psi_1}{\partial t^2} =$$
$$= (-\omega_0^2 \Psi + \frac{\partial^2 \Psi}{\partial t^2} + i2\omega_0 \frac{\partial \Psi}{\partial t})e^{i\omega_0 t} =$$
$$= (-\omega_0^2 \Psi + \frac{\partial^2 \Psi}{\partial t^2} + i2\omega_0 (\vec{v}\nabla\Psi - i\omega_0\Psi))e^{i\omega_0 t} =$$
$$= (\omega_0^2 \Psi + \frac{\partial^2 \Psi}{\partial t^2})e^{i\omega_0 t}.$$

Thus, we obtain the equation

(e.8) $\quad \frac{\partial^2 \Psi}{\partial t^2} = -\omega_0^2 \Psi,$

that is equal to Klein-Gordon equation when $\triangle\psi = 0$, that is, when the momentum $p = 0$, and consequently, the velocity is equal to zero, and with the solution $\Psi(x, y, z, t) = \Phi(x, y, z)e^{-i\omega_0 t}$.

Then we have that it holds the Schrödinger equation for the total energy, $E\Psi = i\hbar\frac{\partial \Psi}{\partial t}$, and the total energy formula (3) introduced in this paper,

$$E = \ |\int i\hbar\frac{\partial \Phi(x-v_x t, y-v_y t, z-v_z t)e^{-i\omega_0 t}}{\partial t}dV|/\mathbf{1}_\Phi =$$
$$= |\int i\hbar\frac{\partial \Phi(x,y,z)e^{-i\omega_0 t}}{\partial t}dV|/\mathbf{1}_\Phi =$$
$$= |\int i\hbar(-i\omega_0)\Phi(x,y,z)e^{-i\omega_0 t}dV|/\mathbf{1}_\Phi =$$
$$= \hbar\omega_0(\int \Phi(x,y,z)dV)/\mathbf{1}_\Phi = \hbar\omega_0,$$

from the fact that for the particles is satisfied the normalization principle (2), $\mathbf{1}_\Phi = \int \Phi(x, y, z)dV$.

$\square$

But, if the velocity of a particle is equal to zero, then we can not have any phenomena of the plain waves, and as we will see, in the interesting cases when $\vec{v}$ is different from zero, the equation (e.4) of propagation of particles is completely different from the equation Klein-Gordon, and consequently, from the Schrödinger equation.

The wave-particle-duality, the fundamental component of the new quantum formalism in Bohrs opinion, was reformulated by incorporating the results of some experiments accomplished in the last decades of twentieth century.

The Bohrs complementarity principle stated the mutual exclusiveness and joint full completeness of the two (classical) descriptions of quantum systems; after Einstein-Podolsky-Rosenpaper, the wave-particle duality, or wave-particle complementarity, could be expressed by stating that it is impossible to build up an experimental arrangement in which we observe at the same time both corpuscular and wave aspects. In a two-slit experiment, they would correspond, respectively, to the which-way knowledge and the observation of interference pattern. Bohr showed this mutual exclusivity in numerous examples [14], and linked it to the unavoidable disturbance inherent in any measurement event.

Not everyone agreed with this interpretation, or with Born and Heisenberg's statement about wave-particle-duality. Einstein and Schrödinger were among the most notable dissenters. Until the ends of their lives they never fully accepted the Copenhagen doctrine. Einstein was dissatisfied with the reliance upon probabilities. But even more fundamentally, he believed that nature exists independently of the experimenter, and the motions of particles are precisely determined. It is the job of the physicist to uncover the laws of nature that govern these motions, which, in the end, will not require statistical theories. The fact that quantum mechanics did seem consistent only with sta-

46

tistical results and could not fully describe every motion was for Einstein an indication that quantum mechanics was still incomplete.

Recently it was demonstrated that intermediate particle-wave behaviors exist and, in addition to that, there are single experiments in which both classical wave-like and particle-like behaviors are showed total and simultaneously on an individual system [8]. For instance, in the Boses double-prism experiment , tunneling and perfect antico-incidence were observed in single photon states.

Consequently, the meaning of the wave-particle duality must incorporate the simultaneous use of the two classical descriptions in the interpretation of experiments, loosing their original mutual exclusivity , which is incorporated as an extreme case in the new interferometric duality, a continuous quantum concept.

In fact, our results demonstrate that Ψ of any elementary particle is always a wavepacket (with a "corpuscular" space-geometric distribution $\Phi(x, y, z, t)$) delimited in the space at each time instance, but that has contemporarily an oscillation in the complex space ('phase wave'), expressed by its complex component $e^{i(-\frac{\vec{p}\cdot\vec{r}}{\hbar} - \omega_0 t)}$.

Now we will show that each relatively dense stream of particles of the same type, velocity, energy and direction of propagation compose a complex plain wave [21], as, for example, an electromagnetic plain wave when these particles are photons.

Proposition 6 *Any dense stream of elementary particles of the same type, velocity, energy and direction of propagation compose a perfect plain wave.*

Proof: Let us suppose a stream of particles that propagates along the x axis, with the distance between two consecutive particles in this stream of particles is $\triangle x \ll \lambda$, where $\lambda = \frac{2\pi\hbar}{p}$ and p is the momentum of each particle.

Then we have that this stream of particles in the moment $t = N\triangle t$, $N \gg 1$ (where $n = N$ is the last emitted particle in the source at $x = 0$, and $n = 0$ is the first emitted particle from this source), is equal to:

$$\sum \Psi_n(x, y, z, t) = \sum_{n=0,1,..,N} \Phi_n(x - v(t - n\triangle t)e^{i(\frac{2\pi}{\lambda}v(t-n\triangle t) - \omega_0(t-n\triangle t))} =$$
$$= \sum_{n=0,1,..,N} \Phi_n(x - (N-n)v\triangle t)e^{i(\frac{2\pi}{\lambda}v(N-n)\triangle t - \omega_0(N-n)\triangle t)} =$$
$$= \sum_{j=N,N-1,..,0} \Phi_n(x - jv\triangle t)e^{i(\frac{2\pi}{\lambda}vj\triangle t - \omega_0(j\triangle t)},$$

Thus in the limit case when $\triangle t \mapsto 0$, we obtain that:

$$lim_{\triangle t \to 0} Re(\sum \Psi_n(x, y, z, t)) = ARe(\int_{\alpha N\triangle t}^0 e^{i\alpha t'} d\alpha t') =$$

(where $\alpha = \frac{2\pi}{\lambda}v - \omega_0$, $t' = j\triangle t$)

$$= A \int_{\alpha N\triangle t}^0 \cos(\alpha t')d\alpha t' =$$
$$= A\sin(\frac{2\pi}{\lambda}vt - \omega_0 t) = \text{(from } t = N\triangle t)$$
$$= A\sin(\frac{2\pi}{\lambda}x - \omega_0 t) = A\sin(\frac{p}{\hbar}x - \omega_0 t),$$

where A is a constant proportional to the density of this stream of particles.

Thus, this stream of particles is represented by the plain wave $A\sin(\frac{2\pi}{\lambda}x - \omega_0 t)$, with wave-length λ and angular frequency ω_0.

$\square$

The non dense streams will compose a kind of amplitude-modulated complex plain waves.

Consequently, it is valid the following lemma:

Lemma 1 *The Schrödinger equation can be obtained by application of the energy conservation over a complex plain wave $Ae^{i(\frac{p}{\hbar}x - \omega_0 t)}$ of a given stream of particles.*

Proof: Let us consider the stream $\psi = Ae^{i(\frac{p}{\hbar}x - \omega_0 t)}$ of the massive particles, where $\omega_0 = \frac{m_0 c^2}{\hbar}$. If we apply the relativistic energy equation for the elementary particles (the stream is composed by a finite number of them) $E^2 = (m_0 c^2)^2 + (cp)^2$ to this plain wave, we obtain the Klein-Gordon equation:

(e.7) $\quad \triangle \psi - \frac{1}{c^2}\frac{\partial^2}{\partial t^2}\psi = (\frac{\omega_0}{c})^2 \psi$,

whose solution can be given by:

(a) $\quad \psi = \phi(x,y,z,t)e^{-i\omega_0 t}$,

with $\triangle \psi = e^{-i\omega_0 t}\triangle \phi$, so that we obtain the first and second time derivation:

$\frac{\partial \psi}{\partial t} = (-i\omega_0 \phi + \frac{\partial \phi}{\partial t})e^{-i\omega_0 t}$,

$\frac{\partial^2 \psi}{\partial t^2} = (-\omega_0^2 \phi - i2\omega_0 \frac{\partial \phi}{\partial t})e^{-i\omega_0 t}$,

by assuming that $\frac{\partial^2 \phi}{\partial t^2}$ is infinitesimal, that is, when $\beta = \frac{v}{c} \ll 1$. Then, if we substitute this last equation into (e.7), we obtain the non-relativistic (when $\beta = \frac{v}{c} \ll 1$) Schrödinger equation in the absence of potential [10]:

(e.9) $\quad \triangle \phi + i\frac{2m_0}{\hbar}\frac{\partial}{\partial t}\phi = 0$.

In this non-relativistic case, when $v \ll c$, we have that total energy of particles is $E = mc^2 = \frac{m_0 c^2}{\sqrt{1-\beta^2}} \approx m_0 c^2 + \frac{m_0 v^2}{2}$, so that $\omega = \frac{E}{\hbar} = \frac{\omega_0}{\sqrt{1-\beta^2}} \approx \omega_0 + \frac{m_0 v^2}{2\hbar} \approx \omega_0$. Consequently, we obtain that the solution (a) is approximately equal to the plain wave of the coherent stream of massive particles, $\psi = Ae^{i(\frac{p}{\hbar}x - \omega_0 t)} \approx Ae^{i(\frac{p}{\hbar}x - \omega t)} = \phi(x,y,z,t)e^{-i\omega_0 t}$, so that

(b) $\quad \phi(x,y,z,t) = Ae^{i(\frac{p}{\hbar}x - \frac{m_0 v^2}{2\hbar}t)} = Ae^{i(\frac{p}{\hbar}x - \frac{E_k}{\hbar}t)}$

$= iA\sin(\frac{p}{\hbar}x)e^{-i\frac{E_k}{\hbar}t} + A\cos(\frac{p}{\hbar}x)e^{-i\frac{E_k}{\hbar}t}$,

is a "plain-wave" solution of the Schrödinger equation (e.9) in the absence of potential, where $E_k = \frac{m_0 v^2}{2}$ is a non-relativistic kinetic energy of any coherent particle in this plain wave of the stream of particles.

In fact, by substituting $\phi(x,y,z,t) = Ae^{i(\frac{p}{\hbar}x - \frac{E_k}{\hbar}t)}$ from (b) into (e,9), we obtain that $\triangle \phi = -A\frac{p^2}{\hbar^2}e^{i(\frac{p}{\hbar}x - \frac{E_k}{\hbar}t)}$, while, $\frac{\partial}{\partial t}\phi = -iA\frac{E_k}{\hbar}e^{i(\frac{p}{\hbar}x - \frac{E_k}{\hbar}t)} = -iA\frac{m_0 v^2}{2\hbar}e^{i(\frac{p}{\hbar}x - \frac{E_k}{\hbar}t)}$, thus

$\triangle \phi + i\frac{2m_0}{\hbar}\frac{\partial}{\partial t}\phi = (-A\frac{p^2}{\hbar^2} + A\frac{m_0^2 v^2}{\hbar^2})e^{i(\frac{p}{\hbar}x - \frac{E_k}{\hbar}t)} = 0 \cdot e^{i(\frac{p}{\hbar}x - \frac{E_k}{\hbar}t)} = 0$, from the fact that in this non-relativistic case, $p = mv = \frac{m_0 v}{\sqrt{1-\beta^2}} \approx m_0 v$. Clearly, both components, of the general solution (b), $A\sin(\frac{p}{\hbar}x)e^{-i\frac{E_k}{\hbar}t}$ and $A\cos(\frac{p}{\hbar}x)e^{-i\frac{E_k}{\hbar}t}$, are the solutions of (e.9), so that we can use, for example the component solution $\phi(x,y,z,t) = A\sin(\frac{p}{\hbar}x)e^{-i\frac{E_k}{\hbar}t}$ of the Schrödinger equation (e.9) in the absence of potential, used in the case of "particle in box" model (a particle free to move in a small space surrounded by impenetrable barriers). Here we have shown that this model is not a model for a movement of a *single* electron, but *statistical effect* of movement of a stream of electrons, which compose a plain wave $\psi = Ae^{i(\frac{p}{\hbar}x - \omega_0 t)}$, with the same moment $\vec{p} = pe_1 + 0e_2 + 0e_3$ and velocity $\vec{v} = ve_1 + 0e_2 + 0e_3$, with $v \ll c$.
$\square$

It is easy to see that the Klein-Gordon equation (e.7) is very different from the equations (e.3) and (e.4) for a propagation of a single massive particle, if its velocity v is different from zero.

In order to reduce the stationary case (e.4) to Klein-Gordon equation, and, consequently, to Schrödinger equation, it would be necessary that this particle satisfies the following conditions: to propagate with velocity of light in the vacuum $v = c$, to have $\omega_p = 0$, that is $\omega_0 = \frac{pc}{\hbar}$, and with $-i\frac{\vec{c}}{\hbar}\frac{\partial \vec{p}}{\partial t} = \omega_0^2$ (while $\frac{\partial \vec{v}}{\partial t} = 0$ for the real component of the right part of equation (e.4)), that is impossible.

Thus, Schrödinger equation in any case can not represent the *propagation* of a single particle. Thus, we have to deal with the reasonable doubts about some of the previously obtained results for the *single* particles, that are based on the deductions from the Schrödinger equation. Especially when they result with fundamentally non classical characters of a quantum state of a particle(as for example in the case of the Wigner function for a single photon with *negative* probabilities).

Consequently, the Schrödinger equation obtained by application of the energy conservation over a complex plain wave $Ae^{i(\frac{p}{\hbar}x - \omega_0 t)}$, corresponds to its application not to a single particle but to the complete stream of particles.

Because of that it has only a statistical meaning that can be applied to the stream of particles. That is, it is not a wave description of a single particle.

From this point of view, the Schrödinger equation has only a statistical meaning, and a non-determinism is only the consequence of this statistical meaning applicable to single particles. Moreover, the interpretation of module of the solution of Schrödinger equation has not valid support to be considered as a probability to find a particle in a given point of space. The underlying particle's theory, presented in this paper demonstrates that Einstein's idea was correct, and that the differential equations derived previously in this paper demonstrate their deterministic and classical corpuscular nature.

Only the streams of these particles in the determined conditions (the equal energy, momentum and direction of propagation) result in the simple plain waves.

Example 8: Particle in Box.

Let us consider the case of "electron in box" (an electron free to move in a small space $0 \leq x \leq L$ surrounded in $x \leq 0$ and $x \geq L$ by impenetrable potential barriers). It is well known that the solution $\phi(x, y, z, t) = A\sin(\frac{p}{\hbar}x)e^{-i\frac{E_k}{\hbar}t}$ of the Schrödinger equation (e.9) in the absence of potential (see the proof of Lemma above) at barrier $x = L$ has to equal to zero, so that it must hold that $\frac{pL}{\hbar} = n\pi$, for $n = 1, 2, \ldots$. That is, in the possible stationary cases of propagations of electrons along axis x, the kinetic energy (when $v \ll c$) is one of discrete quantum values $E_k = \frac{m_0 v^2}{2} = \frac{p^2}{2m_0} = (\frac{\hbar^2 \pi^2}{2m_0 L^2})n^2 = \frac{1}{2m_0}(\frac{h}{2L})^2 n^2$, for $n = 1, 2, \ldots$

From the fact that each electron in a stream of *coherent* electrons, which compose a plain-wave electrons $\psi = Ae^{i(\frac{p}{\hbar}x - \omega_0 t)}$, has the same momentum and energy, it holds that every single electron must have the quantum kinetic energy obtained above.

Let us show that these results can be obtained without using the statistical meaning expressed by Schrödinger equation, but directly by considering the stationary cases of a propagation of a single electron in this box, based on the partial differential equations which describe the movement of a singe particle, and on general packet-wave solution for them.

As we explained previously, in the *stationary* cases of movements of a single particle, when ω_p and $\omega_1 = \omega_0$ are two constants, we obtain a D'Alambert-like differential

49

equation:

(e.4) $\quad (\frac{v_1}{c})^2 \triangle \Psi_1 - \frac{1}{c^2}\frac{\partial^2 \Psi_1}{\partial t^2} = -\frac{1}{c^2}\frac{\partial \overrightarrow{v_1}}{\partial t}\nabla \Psi_1 - i\frac{\overrightarrow{v_1}}{c^2\hbar}\frac{\partial \overrightarrow{p}}{\partial t}\Psi_1$,

where $\Psi_1 = e^{i\omega_p t}\Psi(\overrightarrow{\mathbf{r}}, t)$ and where $\omega_p = \omega_1 + \frac{\overrightarrow{\mathbf{p}}\,\overrightarrow{v_1}}{\hbar}$, with $\omega_1 = \frac{\partial}{\partial t}(\frac{\overrightarrow{\mathbf{p}}\,\overrightarrow{\mathbf{r}}}{\hbar} + \omega_0 t) = \frac{\overrightarrow{\mathbf{r}}}{\hbar}\frac{\partial \overrightarrow{\mathbf{p}}}{\partial t} + \frac{\partial}{\partial t}(\omega_0 t)$, $v_1 = \sqrt{-\overrightarrow{v_1}\overrightarrow{v_1}}$, are constant, such that $\omega_p = \omega_0 - \frac{pv}{\hbar}$.

In fact we consider that in $0 \leq x \leq L$ this electrons moves with constant velocity $\overrightarrow{\mathbf{v}} = ve_1 + 0e_2 + 0e_3$, and constant moment $\overrightarrow{\mathbf{p}} = m_0 \overrightarrow{\mathbf{v}} = m_0ve_1 + 0e_2 + 0e_3$, so that $\overrightarrow{\mathbf{p}}\,\overrightarrow{\mathbf{v}} = -pv$ (from $e_1 \cdot e_1 = -1$).

The unstationary cases of movements of this particle (which can not be covered by statistical Schrödinger equation), happen only at boundaries, when $-\triangle L \leq x \leq 0$ and $L \leq x \leq L + \triangle L$, where $\triangle L \ll L$ is the place when this particle decelerates to velocity from v to zero and then accelerates to v again. Let us consider the boundary $x = L$ (analog case happens at $x = 0$). In this short period of time we have a constant deceleration $\overrightarrow{\mathbf{a}} = -\frac{\partial \overrightarrow{\mathbf{v}}}{\partial t} = \frac{\partial v}{\partial t}e_1 + 0e_2 + 0e_3 = ae_1 + 0e_2 + 0e_3$. Thus, from $\overrightarrow{v_1} = \frac{\partial}{\partial t}(\overrightarrow{\mathbf{v}}t) = \overrightarrow{\mathbf{v}} + \frac{\partial \overrightarrow{\mathbf{v}}}{\partial t}t = \overrightarrow{\mathbf{v}} - \overrightarrow{\mathbf{a}}t$, it becomes equal to 0 when the velocity of electron becomes zero, so that we obtain $t_0 = \frac{v}{a} \ll \frac{L}{v}$ the short interval of time for full deceleration with $\triangle L = \frac{at_0^2}{2} = \frac{v^2}{2a} \ll \frac{L}{2}$. During this deceleration, this electron looses the kinetic energy, by Compton-effects with bosons of electric field of barrier potential. But after $t \geq t_0$, again by the Compton-effects with bosons of this electric field it has an acceleration in inverse direction so that when it reaches again the position $x = L$ it has the same velocity in opposite direction, and the same kinetic energy during its stationary situation of movement between the barriers.

In the space $0 \leq x \leq L$ this particle is in stationary state, with $\overrightarrow{v_1} = \overrightarrow{\mathbf{v}}$, so that $\frac{\partial \overrightarrow{v_1}}{\partial t} = \frac{\partial \overrightarrow{\mathbf{v}}}{\partial t} = 0$, and $\frac{\partial \overrightarrow{\mathbf{p}}}{\partial t} = 0$, so that we obtain a simpler form of D'Alambert equation:

(e.4.2) $\quad (\frac{v}{c})^2 \triangle \Psi_1 - \frac{1}{c^2}\frac{\partial^2 \Psi_1}{\partial t^2} = 0$,

where $\Psi_1 = e^{i\omega_p t}\Psi(\overrightarrow{\mathbf{r}}, t) = e^{i\omega_p t}(\Phi(\overrightarrow{\mathbf{r}}, t)e^{i(\frac{p}{\hbar}x - \omega_0 t)}) = \Phi(\overrightarrow{\mathbf{r}}, t)e^{i(\frac{p}{\hbar}x - (\omega_0 - \omega_p)t)}$
$= \Phi(\overrightarrow{\mathbf{r}}, t)e^{i\frac{p}{\hbar}(x - vt)}$.

Let us fix a time $t = 0$ when this particle is in position $x = L$. From the fact that in any other next time $t = (\frac{2L}{v})n$, for $n = 1, 2, ...$, it has to be in the same place $x = L$, and with the same phase (condition to be in a stationary stable situation), we must have that $\Psi_1(\overrightarrow{\mathbf{r}}, 0) = \Phi(\overrightarrow{\mathbf{r}}, 0)e^{i(\frac{p}{\hbar}L)} = \Psi_1(\overrightarrow{\mathbf{r}}, \frac{2L}{v}) = \Phi(\overrightarrow{\mathbf{r}}, \frac{2L}{v})e^{i(\frac{p}{\hbar}L - \frac{p}{\hbar}v\frac{2L}{v})}$, and from the fact that geometric form Φ of the wave-packet is equal each time it is in a position $x = L$, we have that $\Phi(\overrightarrow{\mathbf{r}}, 0) = \Phi(\overrightarrow{\mathbf{r}}, \frac{2L}{v})$, so that must hold that $e^{i(\frac{p}{\hbar}L)} = e^{i(\frac{p}{\hbar}L - \frac{p}{\hbar}v\frac{2L}{v})}$, that is, $\frac{p}{\hbar}v\frac{2L}{v} = (2\pi)n$, $n = 1, 2, ...$, and, consequently, we obtain the same solution as from Schrödinger equation, $\frac{pL}{\hbar} = n\pi$, for $n = 1, 2, ...$
$\square$

By this example, we demonstrated that non only we are able to resolve the stationary states for the particles, without using statistical Schrödinger equation, but, moreover, we are able to manage, and explain physically, their unstationary behaviors, as in example above (near to potential barriers). This is one of the main advantages of having new partial differential equations for the movements of a *single* particle in 4-dimensional Minkowsky spaces, instead of using only statistical differential equations of Schrödinger and Klein-Gordon (which are correctly applicable to the coherent streams of particles that obtain a general form of the plain-waves).

This results confirm the fact that a particle may only occupy certain positive energy levels. But it is not true, in current interpretation of Schrödinger equation, that standing wave $sin(\frac{p}{\hbar}x)$, with $sin(\frac{p}{\hbar}L) = 0$, obtained as a component of plain-wave, describes the probability of the position of a single particle at place x, and that a particle may never be detected at certain positions, known as spatial nodes. We demonstrated that an electron wave-packet propagates with constant velocity in this box, so that it has an uniform probability to be at any position in this box, and that its propagation in stationary states corresponds to Newton's laws of classical mechanics (if we consider a point-wise wave-packet of matter of a particle as a kind of "corpuscular" object).

What is really different from classical mechanics, in behavior of the particles, is their behavior during their short-time unstable situations, when a matter can have a mainly perpendicular explosion w.r.t. its trajectory (the points of a propagation of its momentum and energy) near to other matter obstacles. In such a case a wave-packet of a particle explodes instantly (there is no energy transfer which obeys the Einstein's limit for velocity along the trajectory of the particles), from a "point-wave" compact objects into a kind of matter clouds: it is caused by changing of boundary conditions for a wave-packet (w.r.t. a standard space symmetry of the vacuum) and by changing the amplitudes of the harmonics in its Fourier's transformations.

The wrong interpretations of Klein-Gordon equation (and consequently, of Schrödinger equation, which is a non-relativistic approximation of Klein-Gordon equation, as presented in the Proof of Lemma 1), is historically well known, especially the possibility to have the negative probabilities ("indefinite probability density" problem), and negative energy of particles. From 1928 to 1934 these interpretation have been changed different times, from Dirac's assumption that probability density is positive, but are allowed negative energies (contradictory with energy preservation principle), then Pauli and Weiskopf presented a new interpretation of Klein-Gordon equation, as field equation for charged spin-0 field and density is interpreted by charge density. After that Dirac equation acquires a field-theoretic interpretation: it does not longer determine a probability amplitude, rather the field operator for a spin $-\frac{1}{2}$ field.

Hope that this new statistical interpretation, where Klein-Gordon equation is not applied to a single particle, but to a coherent stream of the particles, which forms a plain-wave, avoids all these difficulties of interpretations found in past time. Moreover, the new differential equations for a movement of single elementary particles, with energy which is an integral over the positive density of matter in its wave-packet, returns always a with a positive value for total energy of a particle. Moreover, the complex wave-packet has both properties, its matter density $\Phi(x, y, z, t)$ as a wave-packet which propagates in the space, in the stationary states of this particles obeys the Newton's laws of classical mechanics, while its complex oscillator component $e^{i(-\frac{\vec{p}\cdot\vec{r}}{\hbar} - \omega_0 t)}$, that is an oscillation identical to the complex *plain-wave* (like, for example, the complex electromagnetic plain-wave), is responsible both for the quantum properties of the particles (the discrete energy/momentum levels of particles in their stationary states, in different contexts), and for the wave-like behaviors of single particles as well.

Example 9: non-linear Schrödinger equation.
Let consider the case when a particle with rest-mass m_0 propagates along axis x, with a non-relativistic velocity $v \ll c$, thus by kinetic energy $E_k = \frac{m_0 v^2}{2}$. Then a one-

51

dimensional linear Schrödinger equation,

(1) $\triangle u + i\frac{2m_0}{\hbar}\frac{\partial}{\partial t}u = 0$,

with a general plane-wave solution

$u(x,t) = Ae^{i\theta}$, where $\theta = \frac{p}{\hbar}x - \frac{E_k}{\hbar}t$ is a phase of this wave

$= Ae^{i(\frac{p}{\hbar}x - \frac{m_0v^2}{2\hbar}t)} = Ae^{i(\frac{p}{\hbar}x - \frac{E_k}{\hbar}t)}$

$= iA\sin(\frac{p}{\hbar}x)e^{-i\frac{E_k}{\hbar}t} + A\cos(\frac{p}{\hbar}x)e^{-i\frac{E_k}{\hbar}t}$.

In the case of one-dimensional non-linear Schrödinger equation,

(2) $\frac{\hbar^2}{2m_0}\frac{\partial^2}{\partial x^2}(u|u|) + i\hbar\frac{\partial}{\partial t}u = V(x)u|u|^2$,

the general traveling-wave can not have the constant envelope A, but this envelope has to change with x and time t, so that

(3) $u(x,t) = f(x,t)e^{i\theta}$

where the envelope of this plane-wave is a *real* function (the amplitude) $f(x,t) = |u(x,t)|$, and represents the dependence of density of the plasma (stream of electrons with kinetic energy E_k).

If we consider a one-dimensional particle-in-box problem in $0 < x < L$, the stationary solutions for each single electron has to satisfy the same quantifications of its momentum/energy as in the simple (linear) Schrödinger equation, with $p = m_0v = \frac{\pi\hbar}{L}\cdot n$, for $n = 1, 2, 3, ..$, where m_0 is the rest-mass of an electron. That is with the following possible (constant) velocities, $v = \frac{\pi\hbar}{m_0 L}\cdot n$, for $n = 1, 2, 3, ...$ Consequently, in this system we have for the phase of this traveling-wave, $\theta = \frac{i}{\hbar}(px - E_kt) = \frac{i}{\hbar}(m_0vx - \frac{m_0v^2}{2}t) = i(A_1n - A_2n^2)$ for $n = 1, 2, 3, ..$, where $A_1 = \frac{\pi}{L}\cdot x$ and $A_2 = \frac{\pi^2}{2m_0\hbar L^2}\cdot t$.

From (3) we have that

(4) $\frac{\partial}{\partial t}u = (\frac{\partial f}{\partial t} - i\frac{E_k}{\hbar}f)e^{i\theta}$,

(5) $\frac{\partial^2}{\partial x^2}(u|u|) = \frac{\partial^2}{\partial x^2}(f^2e^{i\theta}) = \frac{\partial}{\partial x}((2f\frac{\partial f}{\partial x} + i\frac{p}{\hbar}f^2)e^{i\theta})$

$= (2f\frac{\partial^2 f}{\partial x^2} + 2(\frac{\partial f}{\partial x})^2 + i\frac{4p}{\hbar}f\frac{\partial f}{\partial x} - (\frac{p}{\hbar})^2f^2)e^{i\theta}$,

so that, by substitution of (4) and (5) in (2) we obtain:

$\frac{\hbar^2}{2m_0}(2f\frac{\partial^2 f}{\partial x^2} + 2(\frac{\partial f}{\partial x})^2 + i\frac{4p}{\hbar}f\frac{\partial f}{\partial x} - (\frac{p}{\hbar})^2f^2) + i\hbar(\frac{\partial f}{\partial t} - i\frac{E_k}{\hbar}f) = V(x)f^3$,

and for $0 < x < L$, (when $V(x) = 0$), we obtain for real and imaginary component of this equation, the following two equations:

(6) $E_kf - \frac{p^2}{2m_0}f^2 + \frac{\hbar^2}{m_0}(\frac{\partial f}{\partial x})^2 + \frac{\hbar^2}{m_0}f\frac{\partial^2 f}{\partial x^2} = 0$, and $\hbar\frac{\partial f}{\partial t} + \frac{4p}{\hbar}f\frac{\partial f}{\partial x} = 0$.

That is, by substitution, $E_k = \frac{m_0v^2}{2}$, $p = m_0v$, we obtain

(7) $\frac{m_0^2v^2}{2\hbar^2}(f - f^2) + (\frac{\partial f}{\partial x})^2 + f\frac{\partial^2 f}{\partial x^2} = 0$, and $\hbar\frac{\partial f}{\partial t} + \frac{4m_0v}{\hbar}f\frac{\partial f}{\partial x} = 0$,

for any discrete value of the velocity $v = \frac{\pi\hbar}{m_0 L}\cdot n$, for $n = 1, 2, ...$

Finally, for each $n = 1, 2, ...$ we obtain a pair of equations

(8) $n^2\frac{\pi^2}{2L^2}(f - f^2) + (\frac{\partial f}{\partial x})^2 + f\frac{\partial^2 f}{\partial x^2} = 0$, and $\hbar\frac{\partial f}{\partial t} + n\frac{4\pi}{L}f\frac{\partial f}{\partial x} = 0$.

We are interested to resolve these equations for a number of distinct consecutive instances of time $0 \le t_i \le T$, where $T = \frac{L}{v} = \frac{m_0}{\pi\hbar n}$ and $0 < x < L$, so that $f(t_i, x)$ can be seen as a function with a parameter (constant) t_i and a unique continues variable x. For a normalization purpose we can consider that the absolute value of real function f is $0 \le |f(t,x)| \le 1$.

$\square$

3.3 Double-slit experiment and derivation of Young's low

In quantum mechanics, the double-slit experiment (often referred to as Young's experiment) demonstrates the inseparability of the wave and particle natures of light and other quantum particles. A coherent light source illuminates a thin plate with two parallel slits cut in it, and the light passing through the slits strikes a screen behind them. The wave nature of light causes the light waves passing through both slits to interfere, creating an interference pattern of bright and dark bands on the screen. However, at the screen, the light is always found to be absorbed as though it were made of discrete particles, that is, photons.

If the light travels from the source to the screen as particles, then on the basis of a classical reasoning, the number that strike any particular point on the screen is expected to be equal to the sum of those that go through the left slit and those that go through the right slit. In other words, according to classical particle physics the brightness at any point should be the sum of the brightness when the right slit is blocked and the brightness when the left slit is blocked. However, it is found that unblocking both slits makes some points on the screen brighter, and other points darker. This can only be explained by the alternately additive and subtractive interference of waves, not the exclusively additive nature of particles, so we know that light must have some particle-wave duality.

Restriction to the two experiments in which either both slits are open or one slit is closed has given rise to the idea of wave-particle complementarity [7, 8] according to which a microscopic object (photon, electron, etc.) would manifest itself as a particle in the which-way experiment but as a wave in the interference experiment. This idea has been felt to be counterintuitive by those not being content with an instrumentalist interpretation of quantum mechanics in which that theory is accepted as just describing phenomena without providing explanations.

The most baffling part of this experiment comes when *only one* photon at a time is fired at the barrier with both slits open. The pattern of interference remains the same, as can be seen if many photons are emitted one at a time and recorded on the same sheet of photographic film. The clear implication is that something with a wavelike nature passes simultaneously through both slits and interferes with itself even though there is only one photon present. (The experiment works with electrons, atoms, and even some molecules too.)

When a laboratory apparatus was developed that could reliably fire single electrons at the screen [15], the emergence of an interference pattern suggested to some authors [3] that each electron was interfering with itself; and, therefore, in some sense the electron had to be going through both slits (that is true from our point of view, based on the instantaneous explosion of the wave-packet Φ of this particle when it passes close to the barrier). For something that most people continue to imagine to be an unimaginably small particle to be able to interfere with itself would suggest that this "sub-atomic particle" was in two places at once, but that idea is strongly at odds with the truism, "You cannot be in two places at the same time," (see principle of contradiction). It was easier to conceptualize the electron as a wave than to accept another, more disturbing implication (from the point-of-view of our everyday notions of reality): that quantum objects are able to exist and behave in ways that defy classical interpretation.

Richard Feynman [17] was fond of saying that all of quantum mechanics can be gleaned from carefully thinking through the implications of this single experiment.

We agree with his point of view, and in the rest we will explain how it works when only one photon at time is fired at the barrier with both slits open. We will show that when a particle passes through one slit, the space symmetry in its propagation is sharply broken so that Φ becomes enormously bigger than in the free vacuum, in the way that it is different from zero in the other slit as well. Thus, our theory demonstrates that is true that "particle wave-packet can be in very short interval of time in both slits". A particle "explosion", described in previous sections, is a phenomena that explains how this particle that passes one of the slits, can "see" other slit as well.

Let us consider the 2-dimensional case, when this particle propagates along the axes x toward the barrier with width $\triangle x$ and with the centers of the slit 1 at point $(0, \frac{d}{2})$ and slit 2 at point $(0, -\frac{d}{2})$. Each particle that propagates along the axis x toward this barrier can have one of the possible interactions with the barrier:

1. The elastic collision with another particle inside this barrier (Compton effect), thus it will be reflected from this barrier.

2. The non-elastic collision, that is, the fusion with another particle inside this barrier, thus it will be absorbed.

3. The non-collision with another particles in this barrier, but in the condition when the previous space symmetry during the propagation is hardly broken. The new boundary condition caused by the presence of this barrier (the particle's distribution of a matter does not overlap with the matter distributions of the particles in the barrier) will produce the non-overlapping instantaneous explosion of this particle along the barrier, thus the particle's distribution will cover contemporarily both distant slits.

We will consider only this third case, when there is no changing of the total energy of a particle. The particle density distribution

$\Phi(\overrightarrow{r} - \overrightarrow{v}t) = \Phi(x - vt, y) = \int \int_{-\infty}^{+\infty} B(k)e^{i(k_x(x-vt)+k_y y)}dk_x dk_y,$

have the amplitudes of the harmonics $B(k)$ that depends on the concrete boundary conditions. Thus, in the space before it encounters the barrier, the values of these harmonics $B(k)$ are established by boundary conditions valid in free (symmetric) space, so that Φ has a distribution in a small volume and with a cylindrically symmetric distribution w.r.t. the direction of propagation, such that the greatest density is in the center ($y = 0$) and it diminishes with the distance from particle's trajectory.

In the moment when the particle comes close to this boundary (approximatively in the center between these two slits), the boundary conditions for the differential equation (e.3) of this particle drastically are changed.

As we explained previously, the sharply changing boundary conditions from the vacuum into the conditions when a particle comes close to this barrier (that is, close to atoms of the barrier), produce an instantaneous explosion of the distribution Φ in this instance of time t, mostly in the direction orthogonal to its direction of propagation, that is along the surface of the barrier.

From the fact that there is no collision with another particles, the total energy E and momentum p of this particle are preserved, thus do not change during its propagation from the front to the back side of the barrier.

Consequently, we have that $p_1 = |\vec{\mathbf{p}_1}| = \hbar\frac{2\pi}{\lambda_1}$ momentum of this particle before the barrier, and $p_2 = |\vec{\mathbf{p}_2}| = \hbar\frac{2\pi}{\lambda_2}$ are equal to the constant momentum p, that is, the wavelength is preserved as well, $\lambda_2 = \lambda_1$. From the fact that the total energy is preserved we have that $E = p_1\vartheta_1 = p_2\vartheta_2$, thus also the phase velocities are preserved, i.e., $\vartheta_2 = \vartheta_1$. Let us show that the velocity of a particle is preserved as well: if it is massive particle it holds directly from the preservation of energy and momentum; for the massless particles we need more investigation, as follows. The medium where propagates this particle is equal on both sides of the barrier (and in the two slits as well), so that we have that for the massless particle its velocity is equal to the group velocity as demonstrated in the end of the previous Section. Thus, we have that $v_2 = |\vec{\mathbf{v}_2}| = \vartheta_2(1 + [\frac{\lambda}{n}\frac{\partial n}{\partial\lambda}]_{\lambda=\lambda_2}) = \vartheta_1(1 + [\frac{\lambda}{n}\frac{\partial n}{\partial\lambda}]_{\lambda=\lambda_1}) = v_1 = |\vec{\mathbf{v}_1}|$.

Consequently, in all cases the values of the velocity of this particles before and after the barrier are equal, so that the barrier with two slits can change only the *direction* of a particle.

Thus, as in the case of the refraction, the total energy E does not change, and the product $\vec{\mathbf{v}}\,\vec{\mathbf{p}}$ as well (so that $\frac{\partial\omega_p}{\partial t} = 0$), and (e.6) is satisfied. Consequently, the differential equation that defines the geometric distribution of Φ of this particle, corresponds to the particular stationary case with an extended D'Alambert equation with $v = v_1 = v_2 > 0$ the constant velocity of propagation of this particle:

(e.4.2) $\quad \triangle\Psi_1 - \frac{1}{v^2}\frac{\partial^2\Psi_1}{\partial t^2} = \Theta(\vec{\mathbf{r}}, t)$.

In the free-space propagation we have that the acceleration $\frac{\partial\vec{\mathbf{v}}}{\partial t} = 0$, so it is different from zero only when a particle is very close to the barrier. But this acceleration is not a result of the forces of external fields (i.e., caused by Compton effects with bosons of these fields), but only a result of internal redistribution of the matter density Φ when it passes both slits contemporarily.

Let us consider the case when this particle is not perfectly on the axis x (with $y = 0$) but propagates parallel to it, with a small distance $\triangle y > 0$. Then, when this particle propagates though both slits, we have that the inertial mass of this particle that propagates in slit 1 is greater than the inertial mass that propagates in slit 2 (in fact, we have that the density distribution of Φ in slit 1 is greater that its density distribution in slit 2, because Φ has a symmetric distribution w.r.t the trajectory at $\triangle y > 0$). This disequilibrium of masses of this particle that passes trough two slits creates a momentum w.r.t. its direction of propagation, and consequently this acceleration $\frac{\partial\vec{\mathbf{v}}}{\partial t} \neq 0$, that changes the direction of the velocity $\vec{\mathbf{v}}$ in a very short interval of time $\triangle t < \frac{\triangle x}{v}$.

After this process, this particle continues to propagate without acceleration in this new direction with angle α w.r.t. the axis x, and again in the free space, so that its space distribution becomes again stable and localized in a very small volume.

Let us consider the situation when this particle arrives at point $x \approx 0$, and when its momentum changes from $\vec{\mathbf{p}_1} = pe_1 + 0e_2 + 0e_3$ into the momentum $\vec{\mathbf{p}_2} = p_{2x}e_1 + p_{2y}e_2 + 0e_3$, in this point $-\frac{\triangle x}{2} \leq x \leq \frac{\triangle x}{2}$. We define the vectors of the positions in slit 1 and 2 by $\vec{\mathbf{r}_1} = (x, \frac{d}{2}, 0)$, $\vec{\mathbf{r}_2} = (x, -\frac{d}{2}, 0)$, and the instance of time t_0

when is concluded the chaining of the particle's momentum. In these two points it must hold the continuity of the phase of its wave-packet $\Psi = \Phi(x,y,z,t)e^{i(-\frac{\vec{p}\cdot\vec{r}}{\hbar}-\omega_0 t)}$, where $\Phi(x,y,z,t)$ drastically changes (in different instances of time during its "explosion") its distribution when it is close to this barrier:

(d.2) $e^{i(-\frac{\vec{p_1}\cdot\vec{r_1}}{\hbar}-\omega_0 t_0)} = e^{i(-\frac{\vec{p_2}\cdot\vec{r_1}}{\hbar}-\omega_0 t_0)}$, $\quad e^{i(-\frac{\vec{p_1}\cdot\vec{r_2}}{\hbar}-\omega_0 t_0)} = e^{i(-\frac{\vec{p_2}\cdot\vec{r_2}}{\hbar}-\omega_0 t_0)}$,

i.e., $e^{i\frac{\vec{p_1}\cdot\vec{r_1}}{\hbar}} = e^{-i\frac{\vec{p_2}\cdot\vec{r_1}}{\hbar}}$, and $e^{i\frac{\vec{p_1}\cdot\vec{r_2}}{\hbar}} = e^{-i\frac{\vec{p_2}\cdot\vec{r_2}}{\hbar}}$, that is, $e^{-i\frac{\vec{p_2}\cdot\vec{r_1}}{\hbar}} = e^{-i\frac{\vec{p_2}\cdot\vec{r_1}}{\hbar}}$.

Consequently, it holds that $-\frac{\vec{p_2}\cdot\vec{r_1}}{\hbar} = 2n\pi - \frac{\vec{p_2}\cdot\vec{r_2}}{\hbar}$, i.e.,

$-\frac{(p_{2x}e_1+p_{2y}e_2+0e_3)(xe_1+\frac{d}{2}e_2+0e_3)}{\hbar} = 2n\pi - \frac{(p_{2x}e_1+p_{2y}e_2+0e_3)(xe_1-\frac{d}{2}e_2+0e_3)}{\hbar}$,

and we obtain that,

(d.3) $p_{2y} = \frac{2\pi\hbar}{d}n$, $\quad$ for $\;n = 0,\pm 1, \pm 2, ...$, thus,

(d.4) $p_{2x} = \sqrt{p^2 - p_{2y}^2} = p\sqrt{1-(\frac{2\pi\hbar}{d}n)^2}$, $\quad$ for $\;n = 0,\pm 1, \pm 2, ...$

so that the angle α, w.r.t. the axis x, of the new direction of this particle after the barrier is obtained by,

(d.5) $\tan\alpha = \frac{p_{2y}}{p_{2x}} = \frac{2\pi\hbar}{d}n/p\sqrt{1-(\frac{2\pi\hbar}{d}n)^2} = \frac{n\lambda}{d}/\sqrt{1-(\frac{n\lambda}{d})^2}$,

where $\lambda = \frac{2\pi\hbar}{p}$ is the wavelength determined by the momentum p of this particle.

In the all practical cases we have that $|n\lambda| << d$, thus we obtain that:

(d.6) $\tan\alpha \approx \frac{n\lambda}{d}$, $\quad$ for $\;n = 0,\pm 1, \pm 2, ...$

that corresponds to the Thomas Young's experiment of two-slit interference, based on his observations of water waves. Consequently, each particle changes its angle of propagation after barrier and, statistically, a stream of these particles that strike a screen behind them creates the Young's interference pattern.

When the particle propagates exactly along the axis x (when $\triangle y = 0$), than in this perfect symmetry the amounts of inertial masses that pass the slits are equally distributed between them, thus we obtain the case when $n = 0$, that is, $\alpha = 0$. To bigger distances $\triangle y < \frac{d}{2}$ of the trajectory of a particle, from the axis x (the perfect spatial symmetry), will correspond the major disequilibriums of the masses that passes through these two slits, so that we will obtain bigger values for the angle of deviation α and, consequently, for the value n. This phenomena is deterministic, so that there is no any probabilistic interpretation for the particle's deviations.

Consequently, the Huygens-Fresnel principles are statistic results that can be obtained from the propagation of a single particle ("corpuscular", with geometric localized (at each fixed instance of time t) distribution $\Phi(x,y,z)$), in the analog way as the statistical result of propagation of particles is a plain wave.

There is a variation of the double-slit experiment in which detectors are placed in either or both of the two slits in an attempt to determine which slit the photon passes through on its way to the screen. Placing a detector even in just one of the slits will result in the disappearance of the interference pattern.

In the presence of detectors, the boundary conditions for the differential equation (e.4.2) in the example above, are very different than without detectors, because in this last case the space symmetry is sharply broken by the plain $x = -\frac{\triangle x}{2}$ of the barrier surface, that makes extensive explosion of the particle's distribution Φ in this plain, so that it covers both slits contemporarily. In the presence of detectors this explosion is diminished by the presence of detectors so that the particle's distribution does not cover both, but only one slit that is more close to this particle. Because of that, the particle will pass only trough one of these two slits, and we will not have the conditions (d.1) of continuity in *both* slits, but only in one of them. Consequently we will not have more the interference Young's pattern.

3.4 The complex wave-packet: What does it mean?

We have seen that generally any particle is determined by the wave-packet $\Psi(x, y, z, t)$ composed by two sub components: by the corpuscular matter distribution $\Phi(x, y, z, t)$ that is a real function, and by the 'phase wave' $e^{i\varphi}$ that is a complex function of the particle's *phase* $\varphi = -\frac{\vec{\mathbf{p}} \cdot \vec{\mathbf{r}}}{\hbar} - \omega_0 t$.

Let us denote by $\vec{\mathbf{r}}_\tau(t) = x(t)e_1 + y(t)e_2 + z(t)e_3$ the vector that lies on the Euclidean 3-dimensional particle's trajectory. The 4-dimensional velocity of a given material point (tangent on its trajectory) in this Minkowski space is then defined by

$\vec{\mathbf{v_4}} = \frac{d\vec{s}}{dt} = ce_0 + v_x e_1 + v_y e_2 + v_z e_3 = ce_0 + \vec{\mathbf{v}}$,

where $\vec{\mathbf{v}} = \frac{\partial}{\partial t} \vec{\mathbf{r}}_\tau(t) = v_x e_1 + v_y e_2 + v_z e_3$ is the standard definition of the velocity in the 3-dimensional Euclidean space, with $v_x = \frac{dx}{dt}, v_y = \frac{dy}{dt}, v_z = \frac{dz}{dt}$ and $v = |\vec{\mathbf{v}}| = \sqrt{v_x^2 + v_y^2 + v_z^2} \leq c$. The unitary 3-dimensional velocity vector is defined by $\vec{i_v} = \frac{\vec{\mathbf{v}}}{v}$.

Then we have that $|\vec{\mathbf{v_4}}| = \sqrt{c^2 - v_x^2 - v_y^2 - v_z^2} = c\sqrt{1 - \beta^2}$, where $\beta = \frac{v}{c}$.

The trajectory of a material point that moves in the Minkowski 4-dimensional space by a velocity $\vec{\mathbf{v_4}}$ is defined by the *unitary tangent* vector of this trajectory $\vec{\tau} = \frac{\vec{\mathbf{v_4}}}{|\vec{\mathbf{v_4}}|} = \frac{\vec{\mathbf{v_4}}}{c\sqrt{1-\beta^2}} = \frac{1}{\sqrt{1-\beta^2}} e_0 + \frac{v_x}{c\sqrt{1-\beta^2}} e_1 + \frac{v_y}{c\sqrt{1-\beta^2}} e_2 + \frac{v_z}{c\sqrt{1-\beta^2}} e_3$.

Then the vector $E_0 \vec{\tau}$, where $E_0 = m_0 c^2$ is the energy of the elementary particle with rest mass $m_0 > 0$, is the well defined *invariant* (independent of a referential system in special relativity theory) 4-dimensional *energy-momentum* vector for massive particles in the Minkowski spaces, that is,

$E_0 \vec{\tau} = mc^2 e_0 + cmv_x e_1 + cmv_y e_2 + cmv_z e_3 = Ee_0 + c\vec{\mathbf{p}}$,

where $m = \frac{m_0}{\sqrt{1-\beta^2}}$ is the relativistic mass for a given velocity v of this particle, $E = mc^2$ is its total relativistic energy, and $\vec{\mathbf{p}} = m\vec{\mathbf{v}}$ is its 3-dimensional momentum. Thus, both fundamental properties of a particle, its energy and its momentum, are two physical values that are propagated on the particle's trajectory with the velocity v. Consequently, the energy and momentum on the particle's trajectory are the functions that depend *only* on the time t.

In the case when $v = c$ (i.e., $\beta = 1$), then all components of this unitary tangent vector of the trajectory $\vec{\tau}$ are infinite, so it is more appropriate to use another nonunitary tangent trajectory vector $\vec{\tau_n} = \sqrt{1 - \beta^2} \vec{\tau} = e_0 + \frac{v_x}{c} e_1 + \frac{v_y}{c} e_2 + \frac{v_z}{c} e_3$, with $0 \leq |\vec{\tau_n}| = \sqrt{1 - \beta^2} \leq 1$.

An 4-dimensional angular wavenumber vector $\vec{\mathbf{k_4}}$, and its correspondent 4-dimensional momentum vector $\vec{\mathbf{p_4}}$, in this four-dimensional time-space are given by

$\vec{\mathbf{k_4}} = k_t e_0 + k_x e_1 + k_y e_2 + k_z e_3$, and $\vec{\mathbf{p_4}} = \hbar \vec{\mathbf{k_4}}$.

The following properties for particle's wave-packets are valid:

Proposition 7 *For the wave-packet that propagates with a given velocity $\vec{v_4}$ we have that the scalar 4-dimensional product, for any $\vec{k_4} = \frac{\omega(k)}{c}e_0 + \vec{k}$, $\vec{k_4}\vec{v_4}$ is independent on the variable spatial angular wavenumber $\vec{k}$, that is $\frac{\partial}{\partial k}(\vec{k_4}\vec{v_4}) = 0$.*
The 4-dimensional momentum of this wave-packet is $\vec{p_4} = \frac{\hbar\omega_0}{c}e_0 + \vec{p}$, where $\vec{p}$ is a standard 3-dimensional momentum of a particle expressed by this 4-dimensional wave-packet.
The phase φ of a wave-packet in a position $\vec{r_4}$ is defined by the following scalar product with this momentum in the Minkowski 4-dimensional space, $\varphi = -\frac{1}{\hbar}\vec{p_4}\vec{r_4}$.

Proof: From the fact that $\vec{k_4}\vec{v_4} = \omega(k) + \vec{k}\,\vec{v} = $ (from (0)) $= \omega_0 + \vec{k_0}\,\vec{v}$, and from the fact that ω_0 and $\vec{k_0} = \frac{\vec{p}}{\hbar}$ does not depend on k, we obtain that $\frac{\partial}{\partial k}(\vec{k_4}\vec{v_4}) = 0$.
It is easy to verify that the 4-dimensional momentum of this wave-packet $\vec{p_4}$ is well defined, and that satisfy the general condition above. In fact, we have that for $\vec{k_4} = \frac{\vec{p_4}}{\hbar} = (\frac{\hbar\omega_0}{c}e_0 + \vec{p})/\hbar = \frac{\omega_0}{c}e_0 + \vec{k_0}$, we satisfy the general condition above, $\vec{k_4}\vec{v_4} = \omega_0 + \vec{k_0}\,\vec{v}$.
□

Notice that the 4-dimensional momentum vector of a particle's wave-packet, $\vec{p_4} = \frac{\hbar\omega_0}{c}e_0 + \vec{p}$, in the time-component of the momentum contains the information of the particle's *energy* $\hbar\omega_0$, while the rest spatial components corresponds to the standard Euclidean momentum of the same particle. Thus, the vector $\vec{E} = c\vec{p_4} = \hbar\omega_0 e_0 + c\,\vec{p}$ is a *variant version* of the well known invariant energy-momentum vector (defined only for the massive particles) $E_0\,\vec{\tau}$, valid for the massless particles as well.
In what follows we will show that the origin of this 4-dimensional momentum vector $\vec{p_4}$ lies on the 4-dimensional particle's trajectory in the Minkowski space (like to the standard 3-dimensional momentum vector $\vec{p}$ whose origin lies on the Euclidean 3-dimensional particle's trajectory).
The energy of the particles defined by (3) is and integral over Euclidean space, so it is only dependent on time in the 4-dimensional Minkowski space, so that $\omega_0 = \frac{E(t)}{\hbar}$ is only dependent on time (for massive particles it is a constant).
Moreover, for the massive particles it was obtained (10) that their momentum p is an integral over Euclidean space, thus only dependent on time.
Consequently, the particle's phase (from Proposition 7) $\varphi(\vec{r_4}) = \varphi(\vec{r}, t) = -\frac{1}{\hbar}\vec{p_4}\vec{r_4} = -\vec{k_0}(t)\,\vec{r} - \omega_0(t)t$ is defined in each point $\vec{r_4} = cte_0 + \vec{r}$ of the Minkowski 4-dimensional space.
We define the gradient of this phase in the 4-dimensional Minkowski space as follows:
$\nabla_4\varphi = \frac{\partial\varphi}{\partial(ct)}e_0 + \frac{\partial\varphi}{\partial x}e_1 + \frac{\partial\varphi}{\partial y}e_2 + \frac{\partial\varphi}{\partial z}e_3 = \frac{1}{c}\frac{\partial\varphi}{\partial t}e_0 + \nabla\varphi$.
In particular we are interested for a transmission of the energy and of the momentum of a particle in the Minkowski space, thus on the trajectory $\vec{r_\tau} = x(t)e_1 + y(t)e_2 + z(t)e_3$ that is a geometric reference for the values of the energy and of the momentum (tangent to the trajectory, as the particles velocity $\vec{v}$).
Consequently, we are interested for the particle's phase on its trajectory, $\varphi_\tau(\vec{r_\tau}, t) = -\vec{k_0}(t)\vec{r_\tau} - \omega_0(t)t = k_{0x}(t)x(t) + k_{0y}(t)y(t) + k_{0z}(t)z(t) - \omega_0(t)t$, with $\frac{\partial\varphi_\tau}{\partial x} = k_{0x}(t)$, $\frac{\partial\varphi_\tau}{\partial y} = k_{0y}(t)$, $\frac{\partial\varphi_\tau}{\partial z} = k_{0z}(t)$, that is, $\nabla\varphi_\tau = \vec{k_0}(t)$.
The 3-dimensional space components of the gradient of the phase on the particles tra-

jectory, for instance $\frac{\partial \varphi_\tau}{\partial x}$ is well defined when $\partial x = v_x dt \neq 0$ (i.e., when $v_x \neq 0$),
so that $\frac{\partial \varphi_\tau}{\partial x} = \frac{\partial \varphi_\tau}{\partial t}\frac{\partial t}{\partial x} = \frac{\partial \varphi_\tau}{\partial t}\frac{1}{v_x}$ (otherwise, if $v_x = 0$ then we impose $\frac{\partial \varphi_\tau}{\partial x} = 0$).
Consequently, they exist when $\frac{\partial \varphi_\tau}{\partial t}$ is well defined, so that the particle's phase on its
trajectory has to be *continuous function* in time. This fact implies the continuous vari-
ation of the momentum $\overrightarrow{\mathbf{p}}(t) = \hbar \overrightarrow{\mathbf{k}_0}(t)$, energy $\hbar\omega_0(t)$ and *trajectory position* $\overrightarrow{\mathbf{r}_\tau}(t)$
as well.
Based on this continuity of the particle's phase φ_τ on its trajectory, the following proposition demonstrates that the trajectory of a particle is determined by the gradient of the
particle's phase φ_τ in the 4-dimensional Minkowski space:

Proposition 8 *The trajectory* $\overrightarrow{\tau_n}(\overrightarrow{r_4})$ *of a wave-packet with a distribution* $\Phi(\overrightarrow{r} - \overrightarrow{v}t)$
in the Minkowski 4-dimensional space is determined by the following equation:
(e.5) $\quad -\nabla_4\varphi_\tau(\overrightarrow{r_4})\overrightarrow{\tau_n}(\overrightarrow{r_4}) = \frac{\omega_0(t)}{c} + \frac{1}{c\hbar}\frac{d\overrightarrow{p_4}(t)}{dt}\overrightarrow{r_4}$,
where $\overrightarrow{r_4} = cte_0 + \overrightarrow{r_\tau}$ *is a point of the trajectory in the Minkowski 4-dimensional space,*
and $\overrightarrow{p_4} = \frac{\hbar\omega_0(t)}{c}e_0 + \overrightarrow{p}(t)$ *is the 4-dimensional momentum of this particle introduced in*
Proposition 7. The constant phase propagation on the particles trajectory corresponds
to the Fermat's principle, with $-\nabla_4\varphi_\tau(t)\overrightarrow{\tau_n}(t) = \frac{\omega_0(t)}{c}$.
The relationship on the particle's trajectory between its velocity and momentum, in the
Minkowski 4-dimensional space, is given by the following equation:
(e.6) $\quad \overrightarrow{v_4}\nabla_4(\overrightarrow{p_4}\overrightarrow{r_4}) - \frac{d\overrightarrow{p_4}}{dt}\overrightarrow{r_4} = \hbar\omega_0$, *or, by elimination of* $\overrightarrow{r_4}$,
(e.7) $\quad \overrightarrow{v_4}(-\nabla_4\varphi_\tau + \frac{\overrightarrow{p_4}}{\hbar}) + \frac{\partial\varphi_\tau}{\partial t} = \omega_0$.

Proof: From the fact that $\overrightarrow{r_\tau}$ is a function of a time t, then the phase on the trajectory
is a function only on time, that is $\varphi_\tau(\overrightarrow{r_\tau}, t) = \varphi_\tau(t)$, with $\frac{d\varphi_\tau}{dt} = \frac{\partial\varphi_\tau}{\partial t}$, we have that,
(e.8) $\quad -\frac{\partial\varphi_\tau}{\partial t} = \omega_0(t) + \overrightarrow{\mathbf{k}_0}(t)\frac{\partial\overrightarrow{r_\tau}}{\partial t} + \overrightarrow{r_\tau}\frac{\partial\overrightarrow{\mathbf{k}_0}(t)}{\partial t} + \frac{\partial\omega_0(t)}{\partial t}t =$
$= \omega_0(t) + \overrightarrow{\mathbf{k}_0}(t)\overrightarrow{v} + \overrightarrow{r_\tau}\frac{d\overrightarrow{\mathbf{k}_0}(t)}{dt} + \frac{d\omega_0(t)}{dt}t = \omega_0(t) + \overrightarrow{\mathbf{k}_0}(t)\overrightarrow{v} + \frac{1}{\hbar}\frac{d\overrightarrow{p_4}(t)}{dt}(cte_0 + \overrightarrow{r_\tau}) =$
$= \omega_0(t) + \overrightarrow{\mathbf{k}_0}(t)\overrightarrow{v} + \frac{1}{\hbar}\frac{d\overrightarrow{p_4}(t)}{dt}\overrightarrow{r_4} = -\frac{d\varphi_\tau}{dt}$, i.e.,
$d\varphi_\tau = -\omega_0(t)dt - \overrightarrow{\mathbf{k}_0}(t)\overrightarrow{v}dt - \frac{1}{\hbar}\frac{d\overrightarrow{p_4}(t)}{dt}\overrightarrow{r_4}dt = -\omega_0(t)dt - \overrightarrow{\mathbf{k}_0}(t)d\overrightarrow{r_\tau} - \frac{1}{\hbar}\frac{d\overrightarrow{p_4}(t)}{dt}\overrightarrow{r_4}dt$,
where $\overrightarrow{r_4} = cte_0 + \overrightarrow{r_\tau}$ is a point of the trajectory in the Minkowski 4-dimensional space,
and $\overrightarrow{p_4} = \frac{\hbar\omega_0(t)}{c}e_0 + \overrightarrow{p}(t)$ is the 4-dimensional momentum of this particle introduced
in Proposition 7, with $\overrightarrow{\mathbf{p}}(t) = \hbar\overrightarrow{\mathbf{k}_0}(t)$.
Consequently, for the points of the trajectory $\overrightarrow{r_4} = cte_0 + \overrightarrow{r_\tau}$ we have that $\nabla\varphi_\tau = \overrightarrow{\mathbf{k}_0}$,
and, consequently,
$-\nabla_4\varphi_\tau(\overrightarrow{r_4})\overrightarrow{\tau_n}(\overrightarrow{r_4}) = (-\frac{1}{c}\frac{\partial\varphi_\tau}{\partial t}e_0 - \overrightarrow{\mathbf{k}_0})(e_0 + \frac{\overrightarrow{v}}{c}) =$
$= \frac{\omega_0(t)}{c} + \frac{1}{c\hbar}\frac{d\overrightarrow{p_4}(t)}{dt}\overrightarrow{r_4}$.
The equation (e.6) is obtained from (e.5) by considering that $-\varphi_\tau = \frac{1}{\hbar}\overrightarrow{p_4}\overrightarrow{r_4}$, and
$\overrightarrow{v_4} = c\overrightarrow{\tau_n}$, while (e.7) is obtained from (e.6) by considering that $-\hbar\frac{\partial\varphi_\tau}{\partial t} = \frac{d(\overrightarrow{p_4}\overrightarrow{r_4})}{dt} = \frac{d\overrightarrow{p_4}}{dt}\overrightarrow{r_4} + \overrightarrow{p_4}\overrightarrow{v_4}$.
Notice that in the case when $\frac{\partial\omega_0}{\partial t} = \frac{\partial\overrightarrow{\mathbf{p}}}{\partial t} = 0$ then we obtain that the projection of the
particle's phase-gradient on its trajectory is constant as well, that is, $-\nabla_4\varphi_\tau\overrightarrow{\tau_n} = \frac{\omega_0}{c}$.
Let us show that it *generally* holds for the phase velocity of a particle (the velocity of
the propagation of the constant phase front),

(e.9) $\vartheta = \frac{\omega_0}{k_0} = \frac{dl}{dt} = |\frac{d\vec{r_\tau}}{dt}|$, where $d\vec{r_\tau} = dl\vec{i_v}$ for the infinitesimal trajectory length dl obtained by this phase velocity in the time-interval dt:

In fact from (e.9) we have that $0 = -\omega_0 dt + k_0 dl = -\omega_0 dt - \vec{k_0}d\vec{r_\tau}$, i.e., that $0 = -\omega_0 - \vec{k_0}\frac{d\vec{r_\tau}}{dt}$. Thus, for this phase velocity we have that φ_τ is constant, that is $-\frac{\partial \varphi_\tau}{\partial t} = 0$, so from (e.8) we obtain that for the phase velocity we have that $\frac{d\vec{p_4}(t)}{dt}\vec{r_4} = 0$, and from (e.5) we obtain $-\nabla_4 \varphi_\tau \vec{\tau_n} = \frac{\omega_0}{c}$.

In this particular constant phase condition we obtain that $-\nabla_4 \varphi_\tau = 0e_0 - \vec{k_0}$, and we obtain that,

$-\nabla_4 \varphi_\tau \vec{\tau_n}cdt = -\vec{k_0}\vec{v}\,dt = k_0 vdt = k_0 dl$.

Then the minimization of $\delta \oint -\nabla_4 \varphi_\tau \vec{\tau_n}cdt = \delta \oint k_0 dl = \delta(\frac{\omega_0}{v}\oint ndl) = \delta \oint ndl$ corresponds to the well known Fermat's principle, where $n = \frac{v}{\vartheta}$ is the refraction index of the medium where this particle propagates.

□

The equation in this proposition demonstrates that the projection of the phase-gradient on the particle's trajectory is a function of the time-changing of the particle's momentum in the Minkowski 4-dimensional space.

Notice that from (e.7) we obtain for the particle's trajectory $0 = \vec{v}(\frac{\vec{p}}{\hbar} - \nabla \varphi_\tau) = -\frac{pv}{\hbar} - \vec{v}\nabla \varphi_\tau$, that is $-\vec{i_v}\nabla \varphi_\tau = \frac{p}{\hbar} = k_0$, i.e., the projection of the phase-gradient $\nabla \varphi_\tau$ on the particles trajectory in the Euclidean 3-dimensional space is equal to the normalized momentum $k_0 = \frac{2\pi}{\lambda}$, where λ is the particle's wave-length (this is direct consequence of the fact that $\nabla \varphi_\tau = \vec{k_0}$).

Notice that in the constant phase propagation case (when we consider the phase velocity of a particle), for the massive particles or for the massless particles whose the total energy does not change, the projection of the phase gradient $\nabla_4 \varphi_\tau(t)$ on the trajectory $\vec{\tau_n}(t)$ in the 4-dimensional Minkowski space is constant.

The determination of the particle's trajectory based on its phase will be presented by the famous double-slit experiments, in the next example:

Example 10: Trajectories in the double-slit experiments.

This example is the continuation of the previous Section, dedicated to the double-slit experiment.

Let us consider that the momentum $\vec{p_1} = pe_1 + 0e_2 + 0e_3$ begin to change its direction at the moment $t = 0$ and position $\vec{r_0} = 0e_1 + 0e_2 + 0e_3$, and finish to change in the moment $t = \triangle t$ and position $\vec{r_1} = \triangle xe_1 + \triangle ye_2 + 0e_3$, with obtained final momentum $\vec{p_2} = p_{2x}e_1 + p_{2y}e_2 + 0e_3$ with the same absolute value $p = \sqrt{p_{2x}^2 + p_{2y}^2}$.

At this point from the continuity of the phase changing and the continuity of the trajectory, it must hold (from (e.5)) that:

(d.7) $Lim_{\varepsilon \to 0}(-\nabla \varphi_\tau(\triangle t - \varepsilon)\vec{\tau}(\triangle t - \varepsilon)) = \frac{\omega_0}{c} + \frac{1}{c\hbar}\frac{d\vec{p}(\triangle t)}{dt}(c\triangle te_0 + \vec{r_1}) =$
$= Lim_{\varepsilon \to 0}(-\nabla \varphi_\tau(\triangle t + \varepsilon)\vec{\tau}(\triangle t + \varepsilon)) = \frac{\omega_0}{c}$, that is,

(d.8) $0 = \frac{1}{c\hbar}\frac{d\vec{p_4}(\triangle t)}{dt}(c\triangle te_0 + \vec{r_1}) = \frac{1}{c\hbar}(0e_0 + \frac{d\vec{p}(\triangle t)}{dt})(c\triangle te_0 + \vec{r_1}) =$
$= \frac{p}{c\hbar}\frac{d\vec{i_v}(\triangle t)}{dt}\vec{r_1} = \frac{p}{c\hbar}\frac{d\vec{i_v}(\triangle t)}{dt}(\triangle xe_1 + \triangle ye_2 + 0e_3)$,

where $0 < \triangle t \approx 0$ and $\triangle x = c\triangle t \approx 0$, that is, $\triangle x << \frac{d}{2}$ where d is a distance between the slits. Because of that we have that $\frac{d\vec{i_v}(\triangle t)}{dt} \approx \frac{1}{\triangle t}(\vec{i_v}(\triangle t) - \vec{i_v}(0)) =$

$\frac{1}{\triangle t}(\sin\alpha e_1 + \cos\alpha e_2 + 0e_3 - (1e_1 + 0e_2 + 0e_3)) = \frac{1}{\triangle t}((\sin\alpha - 1)e_1 + \cos\alpha e_2 + 0e_3)$,

where α is the angle of deviation given by (d.6) of T.Young's experiment of two-slits interference. Consequently, from (d.8) we obtain,

(d.9) $\quad 0 \approx ((\sin\alpha - 1)e_1 + \cos\alpha e_2 + 0e_3)(\triangle x e_1 + \triangle y e_2 + 0e_3)$,

that is, $\triangle y \approx \frac{1-\sin\alpha}{\cos\alpha}\triangle x$, so that $\triangle y \leq \triangle x << \frac{d}{2}$, so that the particles trajectory passes continuously between these two slits.

Differently from the matter distribution of this particle that passes trough both slits, the particle's trajectory (i.e. energy-momentum transfer of this particle) passes through the barrier by continuously changing its direction in this infinitesimal interval of time $\triangle t$. Consequently, from the energy-momentum point of view we obtained the *tunneling effect*.

$\square$

Consequently, the particles phase $\varphi = -\frac{\overrightarrow{\mathbf{p_4}}\overrightarrow{\mathbf{r_4}}}{\hbar} = -\frac{\overrightarrow{\mathbf{p}}\cdot\overrightarrow{\mathbf{r}}}{\hbar} - \omega_0 t$ is responsable for the determination of the particle's trajectory, while the real function $\Phi(\overrightarrow{\mathbf{r_4}}) = \Phi(x, y, z, t)$ represents the particles matter distribution in the Minkowski 4-dimensional space, so that the propagation of any elementary particle is described by the pair of *real* functions in this Minkowski space, (Φ, φ), that, equivalently, can be represented mathematically as a unique *complex* wave-packet $\Psi(\overrightarrow{\mathbf{r_4}}) = \Phi(\overrightarrow{\mathbf{r_4}})e^{i\varphi}$.

Chapter 4

Discussion

4.1 More about ERP paradox

The *physical reality* in ERP is defined as follows (cited from [2]): "*If, without in any way disturbing a system, we can predict with certainty (i.e., with probability equal to unity) the value of a physical quantity, then there exists an element of physical reality corresponding to this physical quantity.*" ERP concludes "that the description of reality as given by a wave function is not complete". In fact, the Schrödinger wave function (2) in [2], $\psi = e^{i\frac{2\pi p_0}{h}x} = e^{i\frac{p_0}{\hbar}x}$, is only a statistical equation (from Proposition 1) for a single particle in a coherent stream of particles (with the same energy, velocity and momentum p_0, that propagates along the axis x), but *exactly* it is an equation of this infinite stream of particles. Let us consider as in [2], that this axes is normalized in the interval $0 \le x \le 1$ (where $x = 0$ corresponds to the source of these coherent particles and $x = 1$ to the infinity.

Let us show that physical reality for ψ is just this coherent infinite stream. In fact, as it was done in [2], for the operator corresponding to the momentum (3) in [2], $p = (i\hbar)\frac{\partial}{\partial x}$, we obtain (4) in [2], $\psi' = p\psi = (i\hbar)\frac{\partial \psi}{\partial x} = p_0\psi$, that is, the momentum for this stream is p_0 (in the standard quantum theory vocabulary: the physical quantity of the operator p has with certainty the value p_0 in a state ψ). It is only the case that it can be applied to each particle in this coherent infinite stream as well, because each particle has the same momentum in it.

By considering that this stream is physical realty for ψ, that is contained only in the full space interval $[0, 1]$, the relative probability (6) in [2], $P(a, b) = \int_a^b \overline{\psi}\psi dx = b - a$ is meaningful for this stream only when $a = 0$ and $b = 1$, thus obtaining the value $b - a = 1$ for (6), confirming the fact that the geometric position of a stream is a physical realty. But if we will try to apply (6) to a single particle of this stream (as it was done in [2]) then we will obtain that for the state ψ the coordinate (when $|b - a| << 1$, thus it is not predicted with certainty) of a particle *is not* a physical quantity. This result is only a confirmation that a state ψ (i.e., Schrödinger equation) is incomplete in order to describe the realty of a *single* particle. It does not mean that "when the momentum of a particle is known (obtained by a direct measurement), its coordinates has no physical reality", because as we have shown in this manuscript, the differential equations (e.2) and (e.3) demonstrate that both of them have contemporarily a physical reality.

We did not consider in any way the de Broglie-Bohm pilot-wave theory, because it is based on the non-locality: the velocity (and the acceleration) of a given particle in their theory does not depend on the local forces as in classical mechanics, and depends on the actual (instantaneous) positions of all particles of the universe. Thus, it depends on the particles that are infinitely far from this particle as well, so that the interaction of them with this given particle can not be based on any kind of bosons that can propagate maximally with the velocity of light in the vacuum. In their theory, it is the wavefunction that carries the faster-than-light effect on the single convoluted particles, and not real bosons. But, how a mathematical (wave)function that is not a physical reality (it propagates the real momentum-energy changes to the particles with the infinite velocity, what is contrary to the special relativity theory), can physically interact with physical realities as are the elementary particles? Bohm treated the wavefunction as a physically *real field*, so that the bosons of this field has to be real physical entities as well (which one?) and to satisfy the limitation on the velocity of propagation as all other particles: what is in contradiction with assumption that it interacts instantaneously with another particles. Consequently, this kind on non-locality is really an unreasonable assumption, and we explained in Section 4 in our theory that a given particle interacts *locally* with another particles.

In the standard quantum formalism, measuring observables is generally thought of as measuring operators on the Hilbert space. For example, measuring position is considered to be a measurement of the position operator. This relationship between physical measurements and Hilbert space operators is, for standard quantum mechanics, an additional axiom of the theory. This theory, by contrast, requires no such measurement axioms (and measurement as such is not a dynamically distinct or special sub-category of physical processes in the theory). In particular, the usual operators-as-observables formalism is, for our new theory, a theorem. A major point of the analysis is that many of the measurements of the observables do not correspond to properties of the particles; they are (as in the case of spin discussed above) measurements of the wavefunction.

The Heisenberg uncertainty principle states that when two complementary measurements are made, there is a limit to the product of their accuracy. As an example, if one measures the position with an accuracy of $\triangle x$, and the momentum with an accuracy of $\triangle x$, then $\triangle p \triangle x \geq h$. If we make further measurements in order to get more information, we disturb the system and change the trajectory into a new one depending on the measurement setup; therefore, the measurement results are still subject to Heisenberg's uncertainty relation. In our new theory, there is always a matter of fact about the position and momentum of a particle. Each particle has a well defined trajectory. Observers have limited knowledge as to what this trajectory is (and thus of the position and momentum). It is the lack knowledge of the particle's trajectory that accounts for the uncertainty relation. What one can know about a particle at any given time is described by the particle's wavefunction differential equation (e.2), (e.3). Since the uncertainty relation can be derived from the wavefunction in other interpretations of quantum mechanics, it can be likewise derived (in the epistemic sense mentioned above), on this theory.

4.2 Conclusions

Entanglement is one of the properties of quantum mechanics that caused Einstein and others to dislike the theory. In 1935, Einstein, Podolsky, and Rosen formulated the EPR paradox, a quantum-mechanical thought experiment with a highly counterintuitive and apparently nonlocal outcome, in response to Niels Bohr's advocacy of the belief that quantum mechanics as a theory was complete [2].

Einstein famously derided entanglement as "spukhafte Fernwirkung" [1] or "spooky action at a distance". It was his belief that future mathematicians would discover that quantum entanglement entailed nothing more or less than an error in their calculations. As he once wrote: "I find the idea quite intolerable that an electron exposed to radiation should choose of its own free will, not only its moment to jump off, but also its direction. In that case, I would rather be a cobbler, or even an employee in a gaming house, than a physicist".

In fact, at the end of this document the authors [2] confirm that Schrödinger wavefunction does not provide a complete description of the physical reality. In the previous sections we explained *why* it happens, and that the reason for it is that Schrödinger equation is only statistically valid for the coordinated stream of particles (with the same energy and momentum), and it is not adequate to describe the propagation of a single particle. Thus, Einstein was right.

The differential equations introduced in this manuscript that describe the geometric properties and propagation of any single particle are absolutely deterministic, that is, the velocity, acceleration, momentum and the energy of each particle are precisely determined in the space and in the time, also if we are not able to measure contemporarily all of them. There is no any probabilistic indeterminateness for them, as assumed by erroneous Copenhagen interpretation. The probabilistic interpretation given to Schrödinger equation is caused by the fact that it is statistical, so if we want to apply it to a single particle, then we can speak only about probabilistic values of this particle (because Schrödinger equation is not an adequate description of a *single* particle). Thus, the probabilistic meaning of the particle's values is not the physical reality, but only a consequence of the statistical nature of Schrödinger equation: there is no any collapse phenomena of the probabilities in the moment of the observation that was the base for consideration of "spooky actions at a distance".

The Bohr and Heisenberg *extension* (Copenhagen interpretation of physical reality) of the probabilistic interpretation of the wavefunction (solution of Schrödinger equation) is not correct. It was only an erroneous assumption based on the fact that quantum mechanics as a theory was incomplete [2].

In the Einstein at all paper [2] in 1935, the authors left open the question of whether or not such a complete description exists, but they believed that such a theory is possible. Why this research was apparently abandoned is an open question for me, just because it was the main motivation for me to try to find this deeper and complete description.

Anyway, I hope that the theory for particles, presented in this manuscript, is a relativistic microcosms counterpart for the Einstein general relativistic theory of macrocosms as well. It is common for all particles and all bosons, as gravitons for example, and can be used for a general unification theory.

Bibliography

[1] A.Einstein. The Born-Einstein letters; correspondence between Albert Einstein and Max and Hedwig Born from 1916 to 1955. *Walker, New York*, 1971.

[2] A.Einstein, B.Podolsky, and N.Rosen. Can quantum-mechanical description of physical reality be considered complete? *Phys. Rev. 47 (10)*, pages 777–780, 1935.

[3] B.Greene. The elegant universe: Superstrings, hidden dimensions, and the quest for the ultimate theory. *Vintage Series, Random House Inc.*, pages 110–111, 2000.

[4] L. Broglie. Recherches sur la theorie des quanta. *Ann. de Phys., 10^e serie, t. III (Janvier-Fervier)*, 1925.

[5] B.Sidharth. The thermodynamic universe. *World Scientific. p. 134. ISBN 9812812342. http://books.google.com/books?id=OUfHR36wSfAC&pg=PA134]*, 2008.

[6] C.Liu, Z.Dutton, C.Behroozi, and L.Hau. Observation of coherent optical information storage in an atomic medium using halted light pulses. *Nature 409 (6819)*, pages 490–493, 2001.

[7] D.Harrison. Complementarity and the Copenhagen interpretation of quantum mechanics. *UPSCALE. Dept. of Physics, U. of Toronto. http://www.upscale.utoronto.ca/GeneralInterest/Harrison/Complementarity/CompCopen.html.*, 2002.

[8] B. Diaz-Pintado and C.Maria. Updating the wave-particle duality. *15th UK and European Meeting on the Foundations of Physics, Leeds, 29-31 March*, 2007.

[9] E.Adelberger, G.Dvali, and A.Gruzinov. Photon mass bound destroyed by vortices. *Phys.Rev.Lett.98:010402*, 2007.

[10] E.Schrödinger. An undulatory theory of the mechanics of atoms and molecules. *Physical Review 28 (6)*, pages 1049–1070, 1926.

[11] H.Minkowski. Raum und zeit. *Jahresberichte der Deutschen Mathematiker-Vereinigung*, pages 75–88, 1909.

[12] J. D. Jackson. Classical electrodynamics. *(3rd ed.). Wiley*, 1998.

[13] L.Hau, S.Harris, Z.Dutton, and C.Behroozi. Light speed reduction to 17 metres per second in an ultracold atomic gas. *Nature 397*, pages 594–598, 1999.

[14] N.Bohr, J.A.Wheeler, and W.H.Zurek. Quantum theory and measurement. *Princeton*, pages 9–49, 1983.

[15] O.Donati, G.F.Missiroli, and G.Pozzi. An experiment on electron interference. *American Journal of Physics 41*, pages 639–644, 1973.

[16] P.Gibs. Is the speed of light constant? *in Carlip, S. Usenet Physics FAQ. University of California, Riverside. http://math.ucr.edu/home/baez/physics/Relativity/SpeedOfLight/speed_of_light.html*, 1997.

[17] R.Feynman, R.Leighton, and M.Sands. The Feynman lectures on physics. *Addison Wesley, Reading, Vol. III*, 1965.

[18] Z.Majkić. Energy and special relativity effects for wave packets in the Minkowski 4-dimensional spaces. *Technical Report 03-10, Compet Research, http://zoranmajkic.webs.com*, 2010.

[19] Z.Majkić. Lagrangian for elementary massless particles in the Minkowski 4-dimensional spaces. *Technical Report 05-10, Compet Research, http://zoranmajkic.webs.com*, 2010.

[20] Z.Majkić. Partial differential equations for wave packets in the Minkowski 4-dimensional spaces. *E-Journal Differential Equations and Control Processes, N.1 February 2011, Publisher: Mathematics and Mechanics Faculty of Saint-Petersburg State University, Russia, http://www.math.spbu.ru/diffjournal/pdf/madjkic.pdf*, 2011.

[21] Z.Majkić. Schrödinger equation and wave packets for elementary particles in the Minkowski spaces. *E-Journal Differential Equations and Control Processes, N.3 July 2011, Publisher: Mathematics and Mechanics Faculty of Saint-Petersburg State University, Russia, http://www.math.spbu.ru/diffjournal/pdf/madjkic2.pdf*, 2011.

Printed by Books on Demand GmbH, Norderstedt / Germany